Unit Price Estimating Methods *Updated* 4th Edition

Editors

John Chiang, PE

Phillip R. Waier, PE, CSI

RSMeans

WILEY

John Wiley & Sons, Inc.

Managing Editor: Mary Greene. Editors: Andrea Sillah and John Chiang. Editorial Assistant: Jessica Deady. Production Manager: Michael Kokernak. Production Coordinator: Wayne Anderson. Composition: Sheryl A. Rose. Proofreader: Jill Goodman. Book and cover design: Norman R. Forgit.

For general information about our other products and services, please contact our Customer Care Department within the United States at (800) 762-2974, outside the United States at (317) 572-3993 or fax (317) 572-4002.

Wiley also publishes its books in a variety of electronic formats. Some content that appears in print may not be available in electronic books. For more information about Wiley products, visit our web site at www.wiley.com.

Library of Congress Cataloging-in-Publication Data:

ISBN: 978-0-87629-016-3

Printed in the United States of America

SKY10046950_050223

Table of Contents

Introduction

Estimates are an important part of everyday life. We estimate the amount of food required to feed our family for a week, or travel time to a desired destination. The preparation of a construction estimate requires the same kind of analysis. In general, we have to determine the amount of materials needed to complete a project, and the time required to install those materials, before assigning a cost to both.

This book, now in its fourth edition, will explore the three interdependent variables that make up an estimate:

- Quantity
- Quality
- Cost

Construction documentation—in the form of plans and specifications—dictates the quality and quantities of materials required. Cost is then determined based on these two elements. If a specific cost/budget must be maintained, then either the quantity or quality of the components is adjusted to meet the cost requirement.

Unit Price Estimating

What is the correct or accurate cost of a given construction project? Is it the total price the owner pays the contractor? Might not another reputable contractor perform the same work for a different cost, whether higher or lower? In fact, there isn't one correct estimated cost for a given project. There are too many variables in construction. At best, the estimator can determine a close approximation of what the final costs will be. The resulting accuracy of the approximation is affected by the amount of detail provided, the amount of time spent on the estimate, and the skill of the estimator. The estimator's skill is based on his or her working knowledge of how buildings are constructed. The estimator's creativity may lead to significant differences between estimates prepared by equally qualified individuals. Using a new method to accomplish a particular work activity (in the form of a new piece of equipment or new methodology, for example) may significantly impact final costs.

The accuracy is determined by the cost accountant at the end of the project. If the project was completed in the allotted time and the profit met the expectation provided in the estimate, then it will be considered accurate. Building construction estimating is clearly not as simple as blindly applying material and labor prices and arriving at a magic figure. The purpose of this text is to make the estimating process easier and more organized for the experienced estimator, and to provide less experienced estimators with a basis for sound estimating practice.

Who Uses Unit Price Estimates?

A unit price estimate is the most detailed and accurate type of construction estimate. It is used by architects, engineers, contractors, and many facilities managers/owners in their daily activities.

Architects/Engineers
Design professionals perform unit price estimates for the following reasons:

- *To evaluate change orders.* When a change order is required, the A/E requests a proposal from the contractor. The proposal is usually supported by a detailed estimate. The A/E may perform a unit price estimate to validate or dispute the costs before approving the change.
- *To verify a schedule of payments.* Periodically, it may be necessary to validate a contractor's schedule of payments (the cost related to various elements of work) for invoicing purposes.

Contractors
Most of the estimates a contractor performs are unit price estimates. Except in the early design stages, owners usually require "hard number" estimates (the price for which the contractor will execute a contract to do the work). Therefore, the contractor must perform the most accurate type of estimate—a unit price estimate. Any binding estimate is usually a unit price estimate.

Facilities Managers/Owners
When facilities managers/owners want an accurate estimate on which to base an authorization for expenditures or loan applications, a unit price estimate may be prepared. Owners (or their representatives) also evaluate contractors' proposals for work or change orders to ongoing projects based on unit price estimates.

When accuracy is required, a unit price estimate is the answer.

About this Book

In addition to collecting and publishing construction cost data for over 60 years, RSMeans has also developed a popular series of estimating and other construction reference books. *Means Unit Price Estimating,* now in its fourth edition, was one of the first—a response to customers who sought information on how to create unit price estimates, division by division, and how to use published cost data most effectively. The book has been used for years by novice and practiced estimators to build and enhance their skills.

This book explains and demonstrates the methods and procedures of unit price estimating, from the plans and specifications all the way to the estimate summary. All aspects of the estimate, including the site visit, the quantity takeoff, pricing, and bidding, are covered in detail. A complete sample estimate is presented to demonstrate the application of these estimating practices.

The book also includes a companion website (**www.rsmeans.com/ supplement/67303B.asp**) where you can access, download, customize, and print the many estimating forms featured in the book.

All construction costs used as examples in this book reflect recent cost data from the annually updated *Means Building Construction Cost Data*—the leading source of thorough and up-to-date costs for the construction industry. Many pages, tables, and charts from the cost guide are reproduced in order to show the origin and development of the data.

The chapters follow the logical progression of the estimating process. Initial chapters include analysis of the plans and specifications, the site visit, and evaluation of collected site data. The quantity takeoff, determination of costs, and final pricing are thoroughly covered, reflecting the importance and required attention to detail for each procedure. Also included are some strategies and principles that may be applied after completion of the estimate.

The final section is devoted to a detailed discussion of estimating for each CSI MasterFormat division, the standard classification system for construction. A sample building project estimate is carried out, by division, to the estimate summary. This process demonstrates the proven estimating techniques described in the book, used by a majority of construction estimators and representing sound estimating practice.

Features of this new fourth edition include the fully updated sample estimates, as well as new updates on computer estimating, which provides an overview of available computer estimating products and their capabilities, from the most basic to quite sophisticated programs. The use of published cost data and its integration into these systems is also covered.

Chapter 1

Estimate Types

There are four basic types of estimates relied on by construction cost estimators. While estimate types may be referred to by different names and may not be always recognized as definitive, most estimators will agree that each has its place in the estimating process. The type of estimate performed is related to the amount of design information available. As a project proceeds through the various stages of design (from schematic design, to design development, to contract documents), the type of estimate changes—and accuracy increases. Figure 1.1 graphically demonstrates the relationship of required time versus resulting accuracy for these four basic estimate types.

1. **Order of Magnitude Estimate:** Loosely described as an educated guess, order of magnitude estimates are also referred to as "napkin estimates," because they are often the result of conversations between contractors (or developers) and clients over dinner, in which an estimate is created on the nearest piece of paper. Order of magnitude estimates can be completed in a matter of minutes. Accuracy is –30% to +50%.

2. **Square Foot and Cubic Foot Estimates:** These are often useful when only the proposed size and use of a planned building is known. Very little information is required. Performing a breakout for this type of estimate enables the designer and estimator to adjust components for the proposed use of the structure (hospital, factory, school, apartments, for example), type of foundation (slab on grade, spread footing, piles), and superstructure (steel, concrete, or a combination), and to focus the cost more closely to the final price. Accuracy is –20% to +30%.

3. **Assemblies (or Systems) Estimate:** This is best used as a budgetary tool in the planning stages of a project. Accuracy is expected at –10% to +20%.

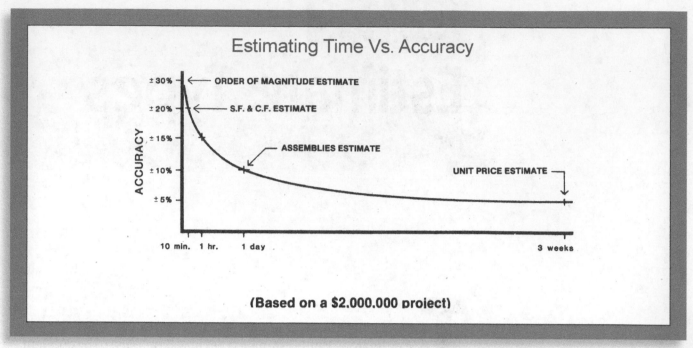

Figure 1.1

4. **Unit Price Estimate:** Working drawings and full specifications are required to complete a unit price estimate—the most accurate of the four types, but also the most time-consuming. Used primarily for bidding purposes, accuracy is –5% to +10%.

Order of Magnitude Estimates

For order of magnitude estimates, the only key information needed is the proposed use and size of the planned structure. The "units" to describe the structure can be very general, and need not be well-defined. What is lacking in accuracy is more than compensated by the minimal time investment—a matter of minutes.

An example:

An office building for a small service company in a suburban industrial park will cost about $840,000.

This type of estimate can be generated after a few minutes of thought, drawing on past experience and comparisons with similar projects. While this rough figure might be appropriate for a project in one region of the country, an adjustment may be required for a change of location and for cost changes over time (price changes, inflation, etc.).

Figure 1.2, from *Means Building Construction Cost Data*, shows examples of a different approach to the order of magnitude estimate, based on unit of use. Note that for some types of buildings, such as hospitals and housing, costs are given per bed or person or per rental unit.

50 17 | Square Foot Costs

		50 17 00 | S.F. Costs	UNIT	UNIT COSTS			% OF TOTAL			
				1/4	MEDIAN	3/4	1/4	MEDIAN	3/4	
41	0010	GARAGES, PARKING	S.F.	31	45.50	78				41
	0020	Total project costs	C.F.	2.92	3.96	5.75				
	2720	Plumbing	S.F.	.88	1.36	2.10	1.72%	2.70%	3.85%	
	2900	Electrical		1.70	2.09	3.28	4.33%	5.20%	6.30%	
	3100	Total: Mechanical & Electrical	↓	3.48	4.85	6.05	7%	8.90%	11.05%	
	3200									
	9000	Per car, total cost	Car	13,100	16,400	21,000				
43	0010	GYMNASIUMS	S.F.	86.50	115	147				43
	0020	Total project costs	C.F.	4.30	5.85	7.15				
	1800	Equipment	S.F.	2.05	3.85	7.40	2.03%	3.30%	5.20%	
	2720	Plumbing		5.45	6.75	8.35	4.95%	6.75%	7.75%	
	2770	Heating, ventilating, air conditioning		5.85	8.95	17.95	5.80%	9.80%	11.10%	
	2900	Electrical		6.55	8.85	11.05	6.60%	8.50%	10.30%	
	3100	Total: Mechanical & Electrical	↓	23.50	33	39	20.50%	24%	29%	
	3500	See also division 11480 (MF2004 11 67 00)								
46	0010	HOSPITALS	S.F.	164	203	300				46
	0020	Total project costs	C.F.	12.55	15.60	22.50				
	1800	Equipment	S.F.	4.20	8.10	13.95	1.10%	2.68%	5%	
	2720	Plumbing		14.25	19.95	25.50	7.60%	9.10%	10.85%	
	2770	Heating, ventilating, air conditioning		21	27	36	7.80%	12.95%	16.65%	
	2900	Electrical		18.05	23.50	36.50	9.85%	11.55%	13.90%	
	3100	Total: Mechanical & Electrical	↓	51	68.50	110	27%	33.50%	37%	
	9000	Per bed or person, total cost	Bed	135,500	217,000	290,000				
	9900	See also division 11700 (MF2004 11 71 00)								
48	0010	HOUSING For the Elderly	S.F.	81.50	103	127				48
	0020	Total project costs	C.F.	5.80	8.05	10.30				
	0100	Site work	S.F.	6.05	8.95	12.90	5.05%	7.90%	12.10%	
	0500	Masonry		2.48	9.25	13.55	1.30%	6.05%	11%	
	1800	Equipment		1.97	2.71	4.32	1.88%	3.23%	4.43%	
	2510	Conveying systems		1.98	2.66	3.61	1.78%	2.20%	2.81%	
	2720	Plumbing		6.05	7.70	9.70	8.15%	9.55%	10.50%	
	2730	Heating, ventilating, air conditioning		3.10	4.40	6.55	3.30%	5.60%	7.25%	
	2900	Electrical		6.05	8.25	10.55	7.30%	8.50%	10.25%	
	3100	Total: Mechanical & Electrical	↓	21	25	33	18.10%	22.50%	29%	
	9000	Per rental unit, total cost	Unit	75,500	88,500	98,500				
	9500	Total: Mechanical & Electrical	"	16,900	19,400	22,600				
50	0010	HOUSING Public (Low Rise)	S.F.	68.50	95	124				50
	0020	Total project costs	C.F.	6.10	7.60	9.45				
	0100	Site work	S.F.	8.70	12.55	20.50	8.35%	11.75%	16.50%	
	1800	Equipment		1.86	3.04	4.63	2.26%	3.03%	4.24%	
	2720	Plumbing		4.94	6.50	8.25	7.15%	9.05%	11.60%	
	2730	Heating, ventilating, air conditioning		2.48	4.81	5.25	4.26%	6.05%	6.45%	
	2900	Electrical		4.14	6.15	8.55	5.10%	6.55%	8.25%	
	3100	Total: Mechanical & Electrical	↓	19.65	25.50	28.50	14.50%	17.55%	26.50%	
	9000	Per apartment, total cost	Apt.	75,000	85,500	107,500				
	9500	Total: Mechanical & Electrical	"	16,000	19,800	21,900				
51	0010	ICE SKATING RINKS	S.F.	58.50	137	150				51
	0020	Total project costs	C.F.	4.30	4.40	5.05				
	2720	Plumbing	S.F.	2.19	4.10	4.19	3.12%	3.23%	5.65%	
	2900	Electrical		6.25	9.60	10.15	6.30%	10.15%	15.05%	
	3100	Total: Mechanical & Electrical	↓	10.45	14.75	18.45	18.95%	18.95%	18.95%	
52	0010	JAILS	S.F.	178	230	297				52
	0020	Total project costs	C.F.	16.05	22.50	27.50				
	1800	Equipment	S.F.	6.95	20.50	35	2.80%	5.55%	11.90%	
	2720	Plumbing		18.15	23	30.50	7%	8.90%	13.35%	
	2770	Heating, ventilating, air conditioning		16.05	21.50	41.50	7.50%	9.45%	17.75%	
	2900	Electrical	↓	18.55	24.50	30.50	8.20%	11.55%	14.70%	

Figure 1.2

Credit: *Means Building Construction Cost Data 2007*

3

Square Foot & Cubic Foot Estimates

Square foot and cubic foot estimates are most appropriate prior to preparing plans or preliminary drawings, when budgetary parameters are being analyzed and established. In Figure 1.2, note that costs for each type of project are presented first as "total project costs" by square foot or cubic foot. These costs are then broken down into different construction components, and then into the relationship of each component to the project as a whole (again in terms of costs per square foot). This breakdown enables the designer, planner, or estimator to adjust certain components according to the unique requirements of the proposed project.

Historical data for square foot costs of new construction is plentiful. However, the best source of square foot costs is the estimator's own cost records for similar projects, adjusted to the parameters of the project in question. While helpful for preparing preliminary budgets, square foot and cubic foot estimates can also be useful as checks against other, more detailed estimates. While slightly more time is required than with order of magnitude estimates, a greater accuracy is achieved due to more specific definition of the project. A square foot estimate is consistent with the amount of design information available at the schematic design phase.

Assemblies (Systems) Estimates

Rising design and construction costs in recent years have made budgeting and cost efficiency increasingly important in the early stages of building projects. Never before has the estimating process had such a crucial role in the initial planning. Unit price estimating, because of the time and detailed information required, is not suited as a budgetary or planning tool. A faster and more cost-effective method for the planning phase of a building project is the assemblies, or systems, estimate, usually prepared when the architect completes the design development plans. While final design details of the building project are not required, estimators should have solid background knowledge of construction materials and methods, building code requirements, design options, and budgetary restrictions.

The systems method is a logical, sequential approach that reflects how a building is constructed. Building construction is organized into seven major components according to ASTM's UNIFORMAT II classification system, *Standard Classification for Building Elements and Related Site Work*.

Assemblies Major Groups:

A – Substructure

B – Shell

C – Interiors

D – Services

E – Equipment & Furnishings

F – Special Construction & Demolition

G – Building Site Work

Each major group is further broken down into systems that incorporate several different items into an assemblage commonly used in building construction. Figure 1.3 is an example of a typical system: drywall partitions/wood stud framing.

It is important to note that in the assemblies format, a construction component may appear in more than one division. For example, concrete is considered part of Division A – Substructure, as well as B – Shell, and G – Building Site Work. Each division may incorporate many different areas of construction, and the labor of different trades.

An advantage of the assemblies estimate is that the estimator or designer can substitute one system for another during design development and quickly determine the cost differential. The owner can then anticipate accurate budgetary requirements before final details and dimensions are established.

Assemblies estimates should not be used as substitutes for unit price estimates. While the systems approach can be an invaluable tool in the planning stages of a project, it should be supported by unit price estimating when greater accuracy is required.

Unit Price Estimates

The unit price estimate is the most accurate and detailed of the four estimate types, and therefore takes the most time to complete. Detailed working drawings and specifications (contract documents) must be available, and all decisions regarding the building's materials and methods must be made. Because there are fewer variables, the estimate can therefore be more accurate.

Working drawings and specifications are needed to determine the quantities of materials, equipment, and labor. Current and accurate costs for these items (unit prices) are also necessary. While these costs may come from different sources, whenever possible, the estimator should use prices based on experience or cost figures from similar projects. If no records are available, prices may be determined instead from an up-to-date, industry-standard cost data book, such as *Means Building Construction Cost Data*.

Because of the detail involved and the need for accuracy, unit price estimates require a great deal of time and expense to complete properly. For this reason, unit price estimating is best suited for construction bidding. It can also be effective for determining certain detailed costs in conceptual budgets or during design development.

Most construction specification manuals and cost reference books, including *Means Building Construction Cost Data,* divide all unit price information into the 49 MasterFormat™ divisions as adopted by the Construction Specifications Institute.

CONSOLIDATED ESTIMATE

PROJECT		CLASSIFICATION				SHEET NO.	
LOCATION		ARCHITECT				ESTIMATE NO.	
TAKE OFF BY	QUANTITIES BY	PRICES BY	EXTENSIONS BY			DATE	
						CHECKED BY	

DESCRIPTION	NO.	DIMENSIONS	QUANTITIES		MATERIAL		LABOR		EQUIPMENT		TOTAL	
			UNIT		UNIT COST	TOTAL	UNIT COST	TOTAL	UNIT COST	TOTAL	UNIT COST	TOTAL

Figure 3.2

Shortcuts

If approached logically and systematically, there are a number of shortcuts that can help to save time without sacrificing accuracy. Abbreviations simply save the time of writing things out, for example. An abbreviations list, similar to that in Appendix B of this book, might be posted in a conspicuous place, providing a consistent pattern of definitions for use within an office.

Measurements

All dimensions—whether printed, measured, or calculated—that can be used for determining quantities for more than one item should be listed on a separate sheet and posted for easy reference. Posted gross dimensions can also be used to quickly check for order-of-magnitude errors.

Measurements should be converted to decimal equivalents before calculations are performed to extend the quantities. Figure 3.3 serves as a quick reference for such conversions. Whether converting from feet and inches to decimals, or from cubic feet to cubic yards, use good judgment and common sense to determine significant digits. Rounding off, or decreasing the number of significant digits, should be done only when it will not statistically affect the resulting product.

Units for each item should be consistent throughout the whole project—from takeoff to cost control. In this way, the original estimate can be equitably compared to progress and final cost reports more easily. It will be easier to keep track of a job.

Conversion of Inches to Decimal Parts per Foot

	0	1″	2″	3″	4″	5″	6″	7″	8″	9″	10″	11″
0	0	.08	.17	.25	.33	.42	.50	.58	.67	.75	.83	.92
1/8″	.01	.09	.18	.26	.34	.43	.51	.59	.68	.76	.84	.93
1/4″	.02	.10	.19	.27	.35	.44	.52	.60	.69	.77	.85	.94
3/8″	.03	.11	.20	.28	.36	.45	.53	.61	.70	.78	.86	.95
1/2″	.04	.12	.21	.29	.37	.46	.54	.62	.71	.79	.87	.96
5/8″	.05	.14	.22	.30	.39	.47	.55	.64	.72	.80	.89	.97
3/4″	.06	.15	.23	.31	.40	.48	.56	.65	.73	.81	.90	.98
7/8″	.07	.16	.24	.32	.41	.49	.57	.66	.74	.82	.91	.99

Figure 3.3

A Note on Rounding

The estimator must use good judgment to determine when rounding is appropriate. An overall 2% or 3% variation in a competitive market can often be the difference between getting or losing a job, or between profit or no profit. It's important to establish rules for rounding to achieve a consistent level of precision. As a general guideline, it is best to not round numbers until the final summary of quantities. The final summary is also the time to convert units (square feet of paving to square yards, linear feet of lumber to board feet, etc.).

Figure 3.4 illustrates an example of the effects of premature rounding. If all items in a project were similarly rounded up for "safety," total project costs may include an unwarranted, or at least unaccountable, allowance. An extra 5% may be enough to be counted out of the bidding competition.

Accounting for Labor

Be sure to quantify and include "labor only" items that are not shown on plans. Such items may or may not be indicated in the specifications and might include cleanup, special labor for handling materials, etc.

Conclusion

In general, the quantity takeoff should be organized so that information gathered can be used to future advantage. Scheduling can be made easier if items are taken off and listed by construction phase or by floor. Material purchasing will similarly benefit.

Effect of Rounding off Measurements at Quantity Takeoff

	Actual	Rounded off	% Difference
Length of partition	96'-8"	100'	3.4%
Height of partition	11'-9"	12'	2.1%
Total S.F.	1,135.87 S.F.	1,200 S.F.	5.6%
Cost ($3.83/S.F.)	$4,350.38	$4,596.00	5.6%
+ 15% overhead	$5,002.94	$5,285.40	5.6%
+ 10% profit	$5,503.23	$5,813.94	5.6%
Total cost difference (incl. O&P): $310.71			

Figure 3.4

The following list is a summation of the suggestions covered in this chapter—plus a few more guidelines—that will be helpful during the quantity takeoff:

- Use preprinted forms.
- Transfer carefully and make clear notations of sums carried from one sheet to the next.
- List dimensions consistently.
- Use printed dimensions. Otherwise, measure dimensions carefully if the scale is known and accurate.
- Add printed dimensions for a single entry.
- Convert feet and inches to decimal feet.
- Do not round off until the final summary of quantities.
- Mark drawings as quantities are determined.
- Be alert for changes in scale, or notes such as "NTS" (not to scale).
- Include required items that may not appear in the plans and specs.
- Use building symmetry to avoid repetitive takeoffs.

And perhaps the five most important points:

- Write legibly.
- Be organized.
- Use common sense.
- Be consistent.
- Add explanatory notes to the estimate to provide an audit trail.

Chapter 4

Pricing the Estimate

After the quantities have been determined, prices, or unit costs, must be applied to determine the total costs. Depending on the chosen estimating method (and thus the degree of accuracy required) and the level of detail, these unit costs may be direct or bare costs, or may include overhead, profit, or contingencies. In unit price estimating, the unit costs most commonly used are "bare," or "unburdened." Items such as overhead and profit are usually added to the total direct costs on the bottom line, at the time of the estimate summary.

Sources of Cost Information

One of the most difficult aspects of the estimator's job is determining accurate and reliable bare cost data. Sources for such data are varied, but can be categorized in terms of their relative reliability. The most reliable source of any cost information is the accurate, up-to-date, well-kept records of the estimator's own company. There is no better cost for a particular construction item than the *actual* cost to the contractor of that item from another recent job, modified (if necessary) to meet the requirements of the project being estimated.

Bids from responsible subcontractors are the second most reliable source of cost data. Any estimating inaccuracies are essentially absorbed by the subcontractor. A subcontract bid is a known, fixed cost prior to the project. Whether the price is right or wrong does not matter (as long as it is a responsible bid with no gross errors). The bid is what the appropriate portion of the work will cost. The prime contractor does not have to estimate the work of subcontractors, except for possible verification of the quote.

Quotations by vendors for material costs are, for the same reasons, as reliable as subcontract bids. In this case, however, the estimator must apply estimated labor costs. Thus the "installed" price for a particular item may be more variable. Whenever possible, all price quotations from vendors or subcontractors should be obtained in writing. Qualifications and exclusions should be clearly stated. Inclusions should be checked to be sure that they are complete and as specified. One way to ensure that these requirements are met is to prepare a form on which all subcontractors and vendors must submit quotations. This form can ask all appropriate questions and provide a consistent source of information for the estimator.

The above procedures are ideal, but often, in the realistic haste of estimating and bidding, quotations are received orally, in person, by telephone, by fax, or by e-mail. The importance of gathering all pertinent information is heightened because omissions are more likely. A preprinted form, such as the one shown in Figure 4.1, can be extremely useful to ensure that all required information and qualifications are obtained and understood. How often has the subcontractor stated, "I didn't know that I was supposed to include that"? With the help of such forms, the appropriate questions are asked and answered. *(An example of the use of this form is shown in Figure 9.23.)*

If the estimator has no cost records for a particular item and is unable to obtain a quotation, then the next most reliable source of price information is current unit price cost books, such as *Means Building Construction Cost Data*. Means presents all such data in the form of national averages; these figures must be adjusted to local conditions. This procedure will be explained in Chapter 7. In addition to being a source of primary costs, unit price books can be useful as a reference or as a cross-check for verifying costs obtained elsewhere.

Lacking cost information from any of the above-mentioned sources, the estimator may have to rely on data from old books or adjusted records from an old project. While these types of costs may not be very accurate, they are better than the final alternative—guesswork.

No matter which source of cost information is used, the system and sequence of pricing should be the same as that used for the quantity takeoff. This consistent approach should continue through both accounting and cost control during construction of the project.

Types of Costs

Unit price estimates for building construction may be organized according to the 49 divisions (16 divisions are reserved for future expansion) of the CSI MasterFormat™ 2004. Within each division, the components or individual construction items are identified, listed, and priced. This kind of definition and detail is necessary to complete an accurate estimate. In addition, each item can be broken down further into material, labor, and equipment components.

All costs included in a unit price estimate can be divided into two types: *direct* and *indirect*. Direct costs are those directly linked to the physical construction of a project—those costs without which the project could not be completed. The material, labor, and equipment costs just mentioned, as well as subcontract costs, are all direct costs. These may also be referred to as "bare," or "unburdened" costs.

TELEPHONE QUOTATION

	DATE	
PROJECT	TIME	
FIRM QUOTING	PHONE ( )	
ADDRESS	BY	
ITEM QUOTED	RECEIVED BY	

WORK INCLUDED	AMOUNT OF QUOTATION
DELIVERY TIME **TOTAL BID**	

DOES QUOTATION INCLUDE THE FOLLOWING: If ☐ NO is checked, determine the following:

STATE & LOCAL SALES TAXES	☐ YES	☐ NO	MATERIAL VALUE
DELIVERY TO THE JOB SITE	☐ YES	☐ NO	WEIGHT
COMPLETE INSTALLATION	☐ YES	☐ NO	QUANTITY
COMPLETE SECTION AS PER PLANS & SPECIFICATIONS	☐ YES	☐ NO	DESCRIBE BELOW

EXCLUSIONS AND QUALIFICATIONS

ADDENDA ACKNOWLEDGEMENT **TOTAL ADJUSTMENTS**	
ADJUSTED TOTAL BID	

ALTERNATES

ALTERNATE NO.	
ALTERNATE NO.	
ALTERNATE NO.	
ALTERNATE NO.	
ALTERNATE NO.	
ALTERNATE NO.	
ALTERNATE NO.	

Figure 4.1

Annual Main Office Expenses

Salaries

Owner	$ 80,000
Engineer/Estimator	50,000
Assistant Estimator	35,000
Project Manager	60,000
General Superintendent	50,000
Bookkeeper/Office Manager	28,000
Secretary/Receptionist	18,000

Office Worker Benefits

Workers' Compensation ⎤ FICA & Unemployment ⎥ Medical Insurance ⎥ Miscellaneous Benefits ⎦	38% of Salaries	121,980

Physical Plant

Office & Warehouse (Rental)	30,000
Utilities	2,400
Telephone	3,000
Office Equipment	3,500
Office Supplies	2,000
Auto & Truck (4 vehicles)	24,000

Professional Services

Accounting	4,000
Legal	2,000
Advertising	4,000

Miscellaneous

Dues	1,500
Seminars & Travel	2,500
Entertainment & Gifts	3,000
Uncollected Receivables (1%)	$100,000
TOTAL ANNUAL EXPENSES	**$624,880**

Figure 4.8

While the office example used here is feasible within the industry, keep in mind that it is hypothetical and that conditions and costs vary from company to company.

In order for this company to stay in business without losses (profit is not yet a factor), not only must all direct construction costs be paid, but an additional $624,880 must be recovered during the year in order to operate the office. Remember that the anticipated volume is $10 million for the year. Office overhead costs, therefore, will be approximately 6.3% of annual volume for this example. The most common method for recovering these costs is to apply this percentage to each job over the course of the year. The percentage may be applied in two ways:

1. **Office overhead applied as a percentage of total project costs.** This is probably the most commonly used method and is appropriate where material and labor costs are not separated.

2. **Office overhead applied as a percentage of labor costs only.** This method requires that labor and material costs be estimated separately. As a result, material handling charges are also more easily applied.

The second method described above allows for more precision in the estimate. This method assumes that office expenses are more closely related to labor costs than to total project costs. For example, assume that two companies have the same total annual volume. Company A builds projects that are material-intensive (90% materials, 10% labor). Company B builds projects that are very labor-intensive (10% materials, 90% labor). In order to manage the large labor force, the office (and overhead) expense of Company B will be much greater than that of Company A. As a result, the applicable overhead percentage of B is greater than that of A based on equal annual volumes. For argument's sake, the overhead percentage of *total costs* for Company A is 3%, for Company B, 10%. If company A then gets projects that are more labor-intensive, an allowance of 3% becomes too low and costs will not be recovered. Likewise, if Company B starts to build material-intensive projects, 10% will be too high an overhead figure and bids may no longer be competitive. Office overhead may be more precisely recovered if it is figured as a percentage of labor costs, rather than total costs. In order to do this, a company must determine the ratio of material to labor costs from its historical records. In the example of Figure 4.8, assume that for this company, the ratio is 50/50. Total annual labor costs would be anticipated to be $4,437,500, as calculated below. As a percentage of labor, office overhead will be:

Annual Volume	$10,000,000
Anticipated Overhead (6.3%)	−625,000
Anticipated Profit (5%)	−500,000
Total Bare Costs	8,875,000
Labor (50% of Bare Costs)	$ 4,437,500

By applying this overhead percentage (14%) to labor costs, the company is assured of recovering office expenses even if the ratio of material to labor changes significantly.

The estimator must also remember that if volume changes significantly, then the percentage for office overhead should be recalculated for current conditions. The same is true if there are changes in office staff. Remember that salaries are the major portion of office overhead costs. It should be noted that a percentage is commonly applied to material costs for handling, in addition to and regardless of the method of recovering office overhead costs. This percentage is more easily calculated if material costs are estimated and listed separately.

Profit

Determining a fair and reasonable percentage to be included for profit is not an easy task. This responsibility is usually left to the owner or chief estimator. Experience is crucial in anticipating what profit the market will bear. The economic climate, competition, knowledge of the project, and familiarity with the architect or owner all affect the way in which profit is determined. Chapter 6 will show one way to mathematically determine profit margin based on historical bidding information. As with all facets of estimating, experience is the key to success.

Contingencies

Like profit, contingencies can be difficult to quantify. Especially appropriate in preliminary budgets, the addition of a contingency is meant to protect the contractor, as well as to give the owner a realistic estimate of project costs.

A contingency percentage should be based on the number of "unknowns" in a project. This percentage should be inversely proportional to the amount of planning detail that has been done for the project. If complete plans and specifications are supplied, and the estimate is thorough and precise, then there is little need for a contingency. Figure 4.9, from *Means Building Construction Cost Data*, lists suggested contingency percentages that may be added to an estimate based on the stage of planning and development.

Pricing an estimate and rounding up, or "padding," each individual item is, in essence, adding a contingency. *(See Figure 3.4.)* This method can cause problems, however, because the estimator can never be quite sure of what is the actual cost and what is the "padding," or safety margin, for each item. At the summary, the estimator cannot determine exactly how much has been included as a contingency for the whole project. A much more accurate and controllable approach is the precise pricing of the estimate and the addition of one contingency amount at the bottom line.

01 21 Allowances

01 21 16 – Contingency Allowances

01 21 16.50 Contingencies	Crew	Daily Output	Labor-Hours	Unit	Material	2007 Bare Costs Labor	Equipment	Total	Total Incl O&P
0010 **CONTINGENCIES**									
0020 For estimate at conceptual stage				Project					20%
0050 Schematic stage									15%
0100 Preliminary working drawing stage (Design Dev.)									10%
0150 Final working drawing stage									3%

01 21 55 – Job Conditions Allowance

01 21 55.50 Job Conditions

	Crew	Daily Output	Labor-Hours	Unit	Material	2007 Bare Costs Labor	Equipment	Total	Total Incl O&P
0010 **JOB CONDITIONS** Modifications to total									
0020 project cost summaries									
0100 Economic conditions, favorable, deduct				Project				2%	2%
0200 Unfavorable, add								5%	5%
0300 Hoisting conditions, favorable, deduct								2%	2%
0400 Unfavorable, add								5%	5%
0500 General Contractor management, experienced, deduct								2%	2%
0600 Inexperienced, add								10%	10%
0700 Labor availability, surplus, deduct								1%	1%
0800 Shortage, add								10%	10%
0900 Material storage area, available, deduct								1%	1%
1000 Not available, add								2%	2%
1100 Subcontractor availability, surplus, deduct								5%	5%
1200 Shortage, add								12%	12%
1300 Work space, available, deduct								2%	2%
1400 Not available, add								5%	5%

01 21 57 – Overtime Allowance

01 21 57.50 Overtime

	Crew	Daily Output	Labor-Hours	Unit	Material	2007 Bare Costs Labor	Equipment	Total	Total Incl O&P
0010 **OVERTIME** for early completion of projects or where	R012909-90								
0020 labor shortages exist, add to usual labor, up to				Costs		100%			

01 21 61 – Cost Indexes

01 21 61.10 Construction Cost Index

	Crew	Daily Output	Labor-Hours	Unit	Material	2007 Bare Costs Labor	Equipment	Total	Total Incl O&P
0010 **CONSTRUCTION COST INDEX** (Reference) over 930 zip code locations in									
0020 The U.S. and Canada, total bldg cost, min. (Clarksdale, MS)				%					67.20%
0050 Average									100%
0100 Maximum (New York, NY)									130.90%

01 21 61.20 Historical Cost Indexes

	Crew	Daily Output	Labor-Hours	Unit	Material	2007 Bare Costs Labor	Equipment	Total	Total Incl O&P
0010 **HISTORICAL COST INDEXES** (See Reference section)									

01 21 61.30 Labor Index

	Crew	Daily Output	Labor-Hours	Unit	Material	2007 Bare Costs Labor	Equipment	Total	Total Incl O&P
0010 **LABOR INDEX** (Reference) For over 930 zip code locations in									
0020 the U.S. and Canada, minimum (Clarksdale, MS)				%		29.90%			
0050 Average						100%			
0100 Maximum (New York, NY)						164.50%			

01 21 61.50 Material Index

	Crew	Daily Output	Labor-Hours	Unit	Material	2007 Bare Costs Labor	Equipment	Total	Total Incl O&P
0010 **MATERIAL INDEX** (Reference) For over 930 zip code locations in									
0020 the U.S. and Canada, minimum (Elizabethtown, KY)				%	90.70%				
0040 Average					100%				
0060 Maximum (Ketchikan, AK)					141.60%				

01 21 63 – Taxes

01 21 63.10 Taxes

	Crew	Daily Output	Labor-Hours	Unit	Material	2007 Bare Costs Labor	Equipment	Total	Total Incl O&P
0010 **TAXES**	R012909-80								
0020 Sales tax, State, average				%	4.84%				
0050 Maximum	R012909-85				7.25%				

Figure 4.9

Credit: *Means Building Construction Cost Data 2007*

Bonds

Bonding requirements for a project will be specified in Division 1 – General Requirements, and will be included in the construction contract. Various types of bonds may be required. Listed below are a few common types:

Bid Bond: A form of bid security executed by the bidder or principal and by a surety (bonding company) to guarantee that the bidder will enter into a contract within a specified time and furnish any required Performance or Labor and Material Payment bonds.

Completion Bond: Also known as "Construction" or "Contract" bond. The guarantee by a surety that the construction contract will be completed and that it will be clear of all liens and encumbrances.

Labor and Material Payment Bond: The guarantee by a surety to the owner that the contractor will pay for all labor and materials used in the performance of the contract as per the construction documents. The claimants under the bond are those having direct contracts with the contractor or any subcontractor.

Performance and Payment Bond: (1) A guarantee that a contractor will perform a job according to the terms of the contracts. (2) A bond of the contractor in which a surety guarantees to the owner that the work will be performed in accordance with the contract documents. Except where prohibited by statute, the performance bond is frequently combined with the labor and material payment bond. The payment bond guarantees that the contractor will pay all subcontractors and material suppliers. Figure 4.10 shows typical average rates for performance bonds for building and roadway construction projects.

Surety Bond: A legal instrument under which one party agrees to answer to another party for the debt, default, or failure to perform of a third party.

The Paperwork

At the pricing stage of the estimate, there is typically a large amount of paperwork that must be assembled, analyzed, and tabulated. Generally, the information contained in this paperwork is covered by the following major categories:

- Quantity Takeoff sheets for all general contractor items (Figure 3.1)
- Material supplier written quotations
- Material supplier telephone quotations (Figure 4.1)
- Subcontractor written quotations
- Equipment supplier quotations
- Cost Analysis or Consolidated Cost Analysis sheets (Figures 4.11 and 4.12)
- Estimate Summary sheet (Figures 4.13, 4.14, and 4.15)

Performance Bond

This table shows the cost of a Performance Bond for a construction job scheduled to be completed in 12 months. Add 1% of the premium cost per month for jobs requiring more than 12 months to complete. The rates are "standard" rates offered to contractors that the bonding company considers financially sound and capable of doing the work. Preferred rates are offered by some bonding companies based upon financial strength of the contractor. Actual rates vary from contractor to contractor and from bonding company to bonding company. Contractors should prequalify through a bonding agency before submitting a bid on a contract that requires a bond.

Contract Amount	Building Construction Class B Projects			Highways & Bridges					
				Class A New Construction			Class A-1 Highway Resurfacing		
First $ 100,000 bid	$25.00 per M			$15.00 per M			$9.40 per M		
Next 400,000 bid	$ 2,500	plus	$15.00 per M	$ 1,500	plus	$10.00 per M	$ 940	plus	$7.20 per M
Next 2,000,000 bid	8,500	plus	10.00 per M	5,500	plus	7.00 per M	3,820	plus	5.00 per M
Next 2,500,000 bid	28,500	plus	7.50 per M	19,500	plus	5.50 per M	15,820	plus	4.50 per M
Next 2,500,000 bid	47,250	plus	7.00 per M	33,250	plus	5.00 per M	28,320	plus	4.50 per M
Over 7,500,000 bid	64,750	plus	6.00 per M	45,750	plus	4.50 per M	39,570	plus	4.00 per M

Figure 4.10

Credit: *Means Building Construction Cost Data 2007*

A system is needed to efficiently handle this mass of paperwork and to ensure that everything will get transferred (and only once) from the Quantity Takeoff to the Cost Analysis sheets. Some general rules for this procedure are:

- Write on only one side of any document where possible.
- Code each sheet with a large division number in a consistent place, preferably near one of the upper corners.
- Use Telephone Quotation forms for uniformity in recording prices received from any source, not only telephone quotes.
- Document the source of every quantity and price.
- Keep each type of document in its pile (Quantities, Materials, Subcontractors, Equipment) filed in order by division number.
- Keep the entire estimate in one or more compartmented folders.
- When an item is transferred to the Cost Analysis sheet, check it off.
- If gross subcontractor quantities are known, pencil in the resultant unit prices to serve as a guide for future projects.

All subcontract costs should be properly noted and listed separately. These costs contain subcontractors' markups, and will be treated differently from other direct costs when the estimator calculates the general contractor's overhead, profit, and contingency allowance.

After all the unit prices, subcontractor prices, and allowances have been entered on the Cost Analysis sheets, the costs are extended. In making the extensions, ignore the cents column and round all totals to the nearest dollar. In a column of figures, the cents will average out and will not be of consequence. Indeed, for budget-type estimates, the extended figures could be rounded to the nearest $10, or even $100, with the loss of only a small amount of precision. Finally, each subdivision is added and the results checked, preferably by someone other than the person doing the extensions.

It is important to check the larger items for order of magnitude errors. If the total subdivision costs are divided by the building area, the resultant square foot cost figures can be used to quickly pinpoint areas that are out of line with expected square foot costs.

The takeoff and pricing method, as discussed, has been to utilize a Quantity Sheet for the material takeoff *(Figure 3.1 in Chapter 3)*, and to transfer the data to a Cost Analysis form for pricing the material, labor, and subcontractor items. *(See Figure 4.11)*.

An alternative to this method is a consolidation of the takeoff task and pricing on a single form. An example, the Consolidated Estimate form, is shown in Figure 4.12. The same sequences and recommendations for completing the Quantity Sheet and Cost Analysis form are to be followed when using the Consolidated Estimate form to price the estimate.

The Estimate Summary

When the pricing of all direct costs is complete, the estimator has two choices: all further price changes and adjustments can be made on the Cost Analysis or Consolidated Estimate sheets, *or* total costs for each subdivision can be transferred to an Estimate Summary sheet so that all further price changes, until bid time, will be done on one sheet.

Unless the estimate has a limited number of items, it is recommended that costs be transferred to an Estimate Summary sheet. This step should be double-checked since an error of transposition may easily occur. Preprinted forms can be useful. A plain columnar form, however, may suffice.

If a company has certain standard listings that are used repeatedly, it would save valuable time to have a custom Estimate Summary sheet printed with the items that need to be listed. The Estimate Summary in Figures 4.13 and 4.14 is an example of a commonly used form. The printed MasterFormat division and subdivision headings act as a checklist to ensure that all required costs are included. Figure 4.15 is a Condensed Estimate Summary form. Appropriate column headings or categories for any Estimate Summary form are:

1. Material
2. Labor
3. Equipment
4. Subcontract
5. Total

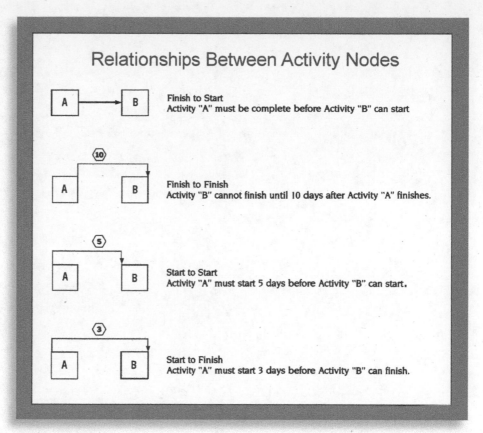

Figure 5.6

Chapter 6

After Completing the Estimate

The goal of most contractors is to make as much money as possible on each job, but more importantly, to maximize return on investment on an annual basis. This means making more money by taking fewer jobs at a higher profit.

Bidding Strategies

One measure of successful bidding is how much money is "left on the table"—the difference between the low bid and next lowest bid. The contractor who consistently takes jobs by a wide margin below the next bidder is obviously not making as much money as possible. Information on competitive public bidding is accessible. Thus, the amount of money left on the table is easily determined and can be the basis for fine-tuning a future bidding strategy.

Since a contractor cannot physically bid every type of job in a geographic area, a selection process must determine which projects should be bid. This process should begin with an analysis of the strengths and weaknesses of the contractor. The following items must be considered as objectively as possible:

- Individual strengths of the company's top management
- Management experience with the type of construction involved, from top management to project superintendents
- Cost records adequate for the appropriate type of construction
- Bonding capability and capacity
- Size of projects with which the company is comfortable
- Geographic area that can be managed effectively
- Unusual corporate assets such as:
  - Specialized equipment availability
  - Reliable and timely cost control systems
  - Strong balance sheet

Market Analysis

Most contractors tend to concentrate on one, or a few, fairly specialized types of construction. From time to time, the company should step back and examine the total picture of the industry they are serving. During this process, the following items should be carefully analyzed:

- Historical trends of the market segment
- Expected future trends of the market segment
- Geographic expectations of the market segment
- Historical and expected competition among other builders
- Risk involved in the particular market segment
- Typical size of projects in this market
- Expected return on investment from the market segment

If several of these areas are experiencing a downturn, then it might be appropriate to examine an alternate market.

Certain steps should be taken to develop a bid strategy for a particular market. The first is to obtain the bid results of jobs in the prospective geographic area. These results should be set up on a tabular basis. This is fairly easy to do in public jobs, since the bid results are normally published (or at least made available) from the agency responsible for the project. In private work, this step can be difficult, since the bid results are not normally divulged by the owner.

For example, assume a public market where all bid prices and the total number of bidders are known. For each "type" of market sector, create a chart showing the percentage left on the table versus the total number of bidders. When the median figure (percent left on the table) for each number of bidders is connected with a smooth curve, the usual shape of the curve is shown in Figure 6.1.

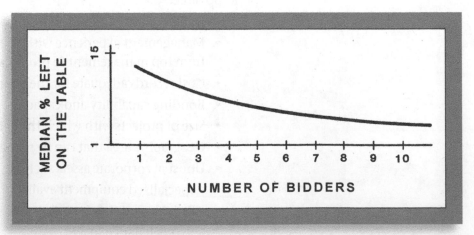

Figure 6.1

The exact magnitude of the amounts left on the table will depend on how much risk is involved with that type of work. If the percentages left on the table are high, then the work can be assumed to be very risky; if the percentages are low, the work is probably not too risky.

Bidding Analysis

If a company has been bidding in a particular market, certain information should be collected and recorded as a basis for a bidding analysis. The percentage left on the table should be tabulated, along with the number of bidders for the projects in the market in which the company was the low bidder. By probability, half of the bids should be above the median line and half below. *(See Figure 6.1.)* If more than half are below the line, the company is doing well; if more than half are above, the bidding strategy should be examined. Once the bidding track record for the company has been established, the next step is to reduce the historical percentage left on the table. One method is to create a chart showing, for instance, the last ten jobs for which the company was the low bidder and the dollar spread between the low and second lowest bid. Next, rank the percentage differences from one to ten (one being the smallest and ten being the largest left on the table). An example is shown in Figure 6.2.

The company's "costs" ($17,170,000) are derived from the company's low bid ($18,887,000) assuming a 10% profit ($1,717,000). The "second bid" is the next lowest bid. The "difference" is the dollar amount between the

Job No.	"Cost"	Low Bid	Second Bid	Difference	% Diff.	% Rank	Profit (Assumed at 10%)
1	$ 918,000	$ 1,009,800	$1,095,000	$ 85,200	9.28	10	$ 91,800
2	1,955,000	2,150,500	2,238,000	87,500	4.48	3	195,500
3	2,141,000	2,355,100	2,493,000	137,900	6.44	6	214,100
4	1,005,000	1,105,500	1,118,000	12,500	1.24	1	100,500
5	2,391,000	2,630,100	2,805,000	174,900	7.31	8	239,100
6	2,782,000	3,060,200	3,188,000	127,800	4.59	4	278,200
7	1,093,000	1,202,300	1,282,000	79,700	7.29	7	109,300
8	832,000	915,200	926,000	10,800	1.30	2	83,200
9	2,372,000	2,609,200	2,745,000	135,800	5.73	5	237,200
10	1,681,000	1,849,100	2,005,000	155,900	9.27	9	168,100
	$17,170,000	$18,887,000		$1,008,000			$1,717,000 = 10% of Cost

Bid Analysis: Comparison of Last 10 Jobs

Figure 6.2

low bid and the second bid. The differences are then ranked based on the percentage of job "costs" left on the table for each.

$$\text{Median \% Difference} = \frac{5.73 + 6.44}{2} = 6.09\%$$

From Figure 6.2, the median percentage left on the table is 6.09%. To maximize the potential returns on a series of competitive bids, a useful formula for pricing profit is needed. The following formula has proven effective:

$$\text{Normal Profit \%} + \frac{\text{Median \% Difference}}{2} = \text{Adjusted Profit \%}$$

$$10.00 + \frac{6.09}{2} = 13.05\%$$

Now apply this adjusted profit percentage to the same list of ten jobs as shown in Figure 6.3. Note that the job "costs" remain the same, but the low bids have been revised. Compare the bottom line results of Figure 6.2 to those of Figure 6.3 based on the two profit margins, 10% and 13.05%, respectively.

Total volume drops from $18,887,000 to $17,333,900.

Net profits rise from $1,717,000 to $2,000,900.

Profits rise while volume drops. If the original volume is maintained or even increased, profits would rise even faster. Note how this occurs. By determining a reasonable increase in profit margin, the company has,

Revised Bid Analysis: Adjusted Profit

Job No.	Company's "Cost"	Revised Low Bid	Second Bid	Adj. Diff.	Profit [10% + 3.05%]		Total
1	$ 918,000	$ 1,037,800	$1,095,000	$ 57,200	$ 91,800 +	$28,000	$ 119,800
2	1,955,000	2,210,100	2,238,000	27,900	195,500 +	59,600	225,100
3	2,141,000	2,420,400	2,493,000	72,600	214,100 +	65,300	279,400
4	(1,005,000)	(1,136,100)	1,118,000 (L)	–	100,500 +	30,600	0
5	2,391,000	2,703,000	2,805,000	102,000	239,100 +	72,900	312,000
6	2,782,000	3,145,100	3,188,000	42,900	278,200 +	84,900	363,100
7	1,093,000	1,235,600	1,282,000	46,400	109,300 +	33,300	142,600
8	(832,000)	(940,600)	926,000 (L)	–	83,200 +	25,400	0
9	2,372,000	2,681,500	2,745,000	63,500	237,200 +	72,300	309,500
10	1,681,000	1,900,400	2,005,000	104,600	168,100 +	51,300	219,400
	$15,333,000	$17,333,900		$517,100			$2,000,900

Figure 6.3

in effect, raised all bids. By doing so, the company loses two jobs to the second bidder (Jobs 4 and 8 in Figure 6.3).

A positive effect of this volume loss is reduced exposure to risk. Since the profit margin is higher, the remaining eight jobs collectively produce more profit than the ten jobs based on the original, lower profit margin. From where did this money come? The money "left on the table" has been reduced from $1,008,000 to $517,100. The whole purpose is to systematically lessen the dollar amount difference between the low bid and the second low bid. This is a hypothetical approach based on a number of assumptions:

- Bidding must be done within the same market in which data for the analysis was gathered.
- Economic conditions should be stable from the time the data is gathered until the analysis is used in bidding. If conditions change, use of such an analysis should be reviewed.
- Each contractor must make roughly the same number of bidding mistakes. For higher numbers of jobs in the sample, this requirement becomes more probable.
- The company must bid additional jobs if the total annual volume is to be maintained or increased. Likewise, if net total profit margin is to remain constant, fewer jobs need be bid.

Even though the accuracy of this strategy is based on these criteria, the concept is valid and can be applied, with appropriate and reasonable judgment, to many bidding situations.

Cost Control & Analysis

An internal accounting system should be used by the contractor to logically allocate and track the gathered costs of a construction project. With this information, a cost analysis can be made about each cost center. The cost centers of a project are the major items of construction (e.g., concrete) that can be subdivided by component. This subdivision should coincide with the system and methods of the quantity takeoff. The major purposes of cost control and analysis are:

- To provide management with a system to monitor costs and progress
- To provide cost feedback to the estimator(s)
- To determine the costs of change orders
- To be used as a basis for submitting payment requisitions to the owner or his representative
- To manage cash flow

A cost control system should be established that is uniform throughout the company and from job to job. The various costs are then consistently allocated. The cost control system might be simplified with a code that provides a different designation for each part of a component cost. The following information should be recorded for each component cost:

- Labor charges in dollars and labor-hours, summarized from weekly time cards, are distributed by code.
- Equipment rental costs are derived from purchase orders or from weekly charges from an equipment company.
- Material charges are determined from purchase orders.
- Appropriate subcontractor charges are allocated.
- Job overhead items may be listed separately or by component.
- Quantities completed to date must also be recorded in order to determine unit costs.

Each component of costs—labor, materials, and equipment—is now calculated on a unit basis by dividing the quantity to date (percentage complete) into the component cost to date. This procedure establishes the actual unit costs to date. The remaining quantities of each component to be completed should be estimated at a unit cost approximating the costs to date. The actual costs to date and the predicted costs are added together to represent the anticipated costs at the end of the project. Typical forms that may be used to develop a cost control system are shown in Figures 6.4 to 6.7.

The analysis of cost centers serves as a useful management tool, providing information on a current basis. Immediate attention is attracted to any center that is operating at a loss. Management can concentrate on this item in an attempt to make it profitable or to minimize the expected loss.

The estimating department can use the unit costs developed in the field as background information for future bidding purposes. Particularly useful are unit labor costs and unit labor-hours (productivity) for the separate components. Current unit labor-hours and labor costs should be integrated into the accumulated historical data.

Frequently, items are added to or deleted from the contract. Accurate cost records are an excellent basis for determining cost changes that result from change orders and requests for additional work.

In order to calculate unit costs, completed quantities need to be determined in the cost records. These calculations are used to determine the percent completion of each cost center. This percentage is used to calculate the billing of completed items for payment requisitions.

A cost system is only as good as the people responsible for coding and recording the required information. Simplicity is the key word. Do not try to break down the code into very small items unless there is a specific need. Continuous cost updating is important so that operations that are not in control can be immediately brought to the attention of management.

PERCENTAGE COMPLETE ANALYSIS

PAGE _____

PROJECT _____ DATE _____

ARCHITECT _____ BY _____ FROM _____ TO _____

NO.	DESCRIPTION	ACTUAL OR ESTIMATED	TOTAL PROJECT	THIS PERIOD		PERCENT TOTAL TO DATE										
				QUANTITY	%	QUANTITY	10	20	30	40	50	60	70	80	90	100
		ACTUAL														
		ESTIMATED														
		ACTUAL														
		ESTIMATED														
		ACTUAL														
		ESTIMATED														
		ACTUAL														
		ESTIMATED														
		ACTUAL														
		ESTIMATED														
		ACTUAL														
		ESTIMATED														
		ACTUAL														
		ESTIMATED														
		ACTUAL														
		ESTIMATED														
		ACTUAL														
		ESTIMATED														
		ACTUAL														
		ESTIMATED														
		ACTUAL														
		ESTIMATED														
		ACTUAL														
		ESTIMATED														
		ACTUAL														
		ESTIMATED														
		ACTUAL														
		ESTIMATED														
		ACTUAL														
		ESTIMATED														
		ACTUAL														
		ESTIMATED														
		ACTUAL														
		ESTIMATED														
		ACTUAL														
		ESTIMATED														
		ACTUAL														
		ESTIMATED														
		ACTUAL														
		ESTIMATED														

Figure 6.4

JOB
PROGRESS REPORT

SHEET _____ OF _____

PROJECT

JOB NO.

LOCATION

YEARS

WORK ITEM	QUANTITY, $, OR %	BEGINNING BALANCE	MONTHS										ENDING BALANCE
	DATE												
	YEAR												

Figure 6.5

MATERIAL COST RECORD

SHEET NO. _____

DATE FROM _____

PROJECT _____

DATE TO _____

LOCATION _____

BY _____

DATE	NUMBER	VENDOR/DESCRIPTION	QTY.	UNIT PRICE						QTY.	UNIT PRICE						QTY.	UNIT PRICE				

Figure 6.6

LABOR COST RECORD

SHEET NO.

DATE FROM:

PROJECT

DATE TO:

LOCATION

BY:

DATE	CHARGE NO.	DESCRIPTION	HOURS	RATE	AMOUNT	HOURS	RATE	AMOUNT	HOURS	RATE	AMOUNT

Figure 6.7

Productivity & Efficiency

When using a cost control system such as the one just described, the derived unit costs should reflect standard practices. Productivity should be based on a five-day, eight-hour-per-day (during daylight hours) work week unless a company's requirements are particularly unique. Installation costs should be derived using normal minimum crew sizes, under normal weather conditions, during the normal construction season. Costs and productivity should also be based on familiar types of construction.

All unusual costs incurred or expected should be separately recorded for each component of work. For example, an overtime situation might occur on every job and in the same proportion. In this case, it would make sense to carry the unit price adjusted for the added cost of premium time. Likewise, unusual weather delays, strike activity, or owner/architect delays should have separate, identifiable cost contributions; these are applied as isolated costs to the activities affected by the delays. This procedure serves two purposes:

- To identify and separate the cost contribution of the delay so job estimates will not automatically include an allowance for these "non-typical" delays.
- To serve as a basis for an extra compensation claim and/or as justification for reasonable extension of the job.

The use of long-term overtime on almost any construction job is counterproductive; that is, the longer the period of overtime, the lower the actual production rate. There have been numerous studies conducted that come up with slightly different numbers, but all have the same conclusion.

As illustrated in Figure 6.8, there can be a difference between the actual payroll cost per hour and the effective cost per hour for overtime work. This is due to the reduced production efficiency with the increase in weekly hours beyond 40. This difference between actual and effective cost is for overtime over a prolonged period. Short-term overtime does not result in as great a reduction in efficiency, and, in such cases, cost premiums would approach the payroll costs rather than the effective hourly costs listed in Figure 6.8. As the total hours per week are increased on a regular basis, more time is lost by absenteeism, and the accident rate increases.

As an example, assume a project where workers are working 6 days a week, 10 hours per day. From Figure 6.8 (based on productivity studies), the actual productive hours are 51.1 hours. This represents a theoretical production efficiency of 51.1/60 or 85.2%.

Depending on the locale, overtime work is paid at time and a half or double time. In both cases, the overall actual payroll cost (including regular and overtime hours) is determined as follows:

04 21 Clay Unit Masonry

04 21 13 – Brick Masonry

04 21 13.13 Brick Veneer Masonry		Crew	Daily Output	Labor-Hours	Unit	Material	2007 Bare Costs Labor	Equipment	Total	Total Incl O&P
2750	Full header every 6th course (7.88/S.F.)	D-8	170	.235	S.F.	11.30	8.05		19.35	25
3000	Jumbo, 6" x 4" x 12" running bond (3.00/S.F.)		435	.092		4.28	3.15		7.43	9.50
3050	Norman, 4" x 2-2/3" x 12" running bond, (4.5/S.F.)		320	.125		4.65	4.29		8.94	11.60
3100	Norwegian, 4" x 3-1/5" x 12" (3.75/S.F.)		375	.107		3.44	3.66		7.10	9.35
3150	Economy, 4" x 4" x 8" (4.50/S.F.)		310	.129		4.09	4.42		8.51	11.25
3200	Engineer, 4" x 3-1/5" x 8" (5.63/S.F.)		260	.154		3.26	5.30		8.56	11.65
3250	Roman, 4" x 2" x 12" (6.00/S.F.)		250	.160		5.40	5.50		10.90	14.30
3300	SCR, 6" x 2-2/3" x 12" (4.50/S.F.)		310	.129		4.92	4.42		9.34	12.15
3350	Utility, 4" x 4" x 12" (3.00/S.F.)	↓	450	.089	↓	3.59	3.05		6.64	8.60
3400	For cavity wall construction, add						15%			
3450	For stacked bond, add						10%			
3500	For interior veneer construction, add						15%			
3550	For curved walls, add						30%			

04 21 13.15 Chimney

		Crew	Daily Output	Labor-Hours	Unit	Material	2007 Bare Costs Labor	Equipment	Total	Total Incl O&P
0010	**CHIMNEY**									
0100	Brick, 16" x 16", 8" flue, scaff. not incl.	D-1	18.20	.879	V.L.F.	18.90	29.50		48.40	65.50
0150	16" x 20" with one 8" x 12" flue		16	1		29	33.50		62.50	83
0200	16" x 24" with two 8" x 8" flues		14	1.143		41.50	38		79.50	104
0250	20" x 20" with one 12" x 12" flue		13.70	1.168		34.50	39		73.50	97.50
0300	20" x 24" with two 8" x 12" flues		12	1.333		47.50	44.50		92	120
0350	20" x 32" with two 12" x 12" flues	↓	10	1.600		60.50	53.50		114	148
1800	Metal, high temp. steel jacket, factory lining, 24" diam.	E-2	65	.862		198	35	24	257	305
1900	60" diameter	"	30	1.867		720	75.50	51.50	847	975
2100	Poured concrete, brick lining, 200' high x 10' diam.					6,525			6,525	7,175
2800	500' x 20' diameter				↓	11,400			11,400	12,600

04 21 13.18 Columns

		Crew	Daily Output	Labor-Hours	Unit	Material	2007 Bare Costs Labor	Equipment	Total	Total Incl O&P
0010	**COLUMNS**	R042110-10								
0050	Brick, 8" x 8", 9 brick per course	D-1	56	.286	V.L.F.	4.94	9.55		14.49	19.95
0100	12" x 8", 13.5 brick		37	.432		7.40	14.40		21.80	30
0200	12" x 12", 20 brick		25	.640		11	21.50		32.50	44.50
0300	16" x 12", 27 brick		19	.842		14.80	28		42.80	59
0400	16" x 16", 36 brick		14	1.143		19.75	38		57.75	79.50
0500	20" x 16", 45 brick		11	1.455		24.50	48.50		73	101
0600	20" x 20", 56 brick		9	1.778		31	59.50		90.50	124
0700	24" x 20", 68 brick		7	2.286		37.50	76		113.50	157
0800	24" x 24", 81 brick		6	2.667		44.50	89		133.50	184
1000	36" x 36", 182 brick	↓	3	5.333		100	178		278	380

04 21 13.30 Oversized Brick

		Crew	Daily Output	Labor-Hours	Unit	Material	2007 Bare Costs Labor	Equipment	Total	Total Incl O&P
0010	**OVERSIZED BRICK**									
0100	Veneer, 4" x 2.25" x 16"	D-8	387	.103	S.F.	4.79	3.54		8.33	10.65
0105	4" x 2.75" x 16"		412	.097		4.33	3.33		7.66	9.80
0110	4" x 4" x 16"		460	.087		4.26	2.98		7.24	9.20
0120	4" x 8" x 16"		533	.075		5.20	2.57		7.77	9.65
0125	Loadbearing, 6" x 4" x 16", grouted and reinforced		387	.103		7.55	3.54		11.09	13.70
0130	8" x 4" x 16", grouted and reinforced		327	.122		8	4.19		12.19	15.20
0135	6" x 8" x 16", grouted and reinforced		440	.091		7.20	3.12		10.32	12.70
0140	8" x 8" x 16", grouted and reinforced		400	.100		8.10	3.43		11.53	14.10
0145	Curtainwall / reinforced veneer, 6" x 4" x 16"		387	.103		11.75	3.54		15.29	18.30
0150	8" x 4" x 16"		327	.122		14.25	4.19		18.44	22
0155	6" x 8" x 16"		440	.091		11.85	3.12		14.97	17.80
0160	8" x 8" x 16"	↓	400	.100		14.35	3.43		17.78	21
0200	For 1 to 3 slots in face, add				↓	25%				

82

Figure 7.1

Figure 7.2 is a graphic representation of how to use the unit price section as presented in *Means Building Construction Cost Data.*

Line Numbers

Every construction item in the Means unit price cost data books has a unique line number that acts as an address so that each item can be quickly located and/or referenced. The numbering system is based on the MasterFormat classification by division. In Figure 7.2, note the bold number in white type, 03 30. This number represents the Means subdivision, in this case, cast-in-place concrete, of the MasterFormat Division 3 – Concrete. All 49 MasterFormat divisions are organized in this manner. Within each subdivision, the data is broken down into major classifications. These major classifications are listed alphabetically and are designated by bold type for both numbers and descriptions.

Each item, or line, is further defined by an individual number. As shown in Figure 7.2, the full line number for each item consists of:

MasterFormat Division – Means subdivision – major classification number – item line number

Each full line number describes a unique construction element. For example, in Figure 7.1, the line number for Norman brick veneer, running bond (per square foot) is 04 21 13.13 3050.

Line Description

Each line has a text description of the item for which costs are listed. In some cases, the description may be self-contained and all-inclusive. Any indented descriptions are delineations (by size, color, material, etc.) or breakdowns of previously described items. An index is provided in the back of *Means Building Construction Cost Data* to aid in locating particular items.

Crew

For each construction element (each line item), a minimum typical crew is designated as appropriate to perform the work. The crew may include one or more trades, foremen, craftsmen, and helpers, and any equipment required for proper installation of the described item. If an individual trade installs the item using only hand tools, the smallest efficient number of tradesmen will be indicated (1 Carp, 2 Carp, etc.). Abbreviations for trades are shown in Figure 7.3. If more than one trade is required to install the item and/or if powered equipment is needed, a crew number will be designated (B-5, D-3, etc.) A complete listing of crews is presented in the reference section in the back of *Means Building Construction Cost Data. (See Figure 7.4.)* Each crew breakdown contains the following components:

1. Number and type of workers designated.
2. Number, size, and type of any equipment required.

How to Use the Unit Price Pages

The following is a detailed explanation of a sample entry in the Unit Price Section. Next to each bold number below is the described item with the appropriate component of the sample entry following in parentheses. Some prices are listed as bare costs; others as costs that include overhead and profit of the installing contractor. In most cases, if the work is to be subcontracted, the general contractor will need to add an additional markup (RSMeans suggests using 10%) to the figures in the column "Total Incl. O&P."

1 Division Number/Title
(03 30/Cast-In-Place Concrete)

Use the Unit Price Section Table of Contents to locate specific items. The sections are classified according to the CSI MasterFormat 2004 system.

2 Line Numbers
(03 30 53.40 3920)

Each unit price line item has been assigned a unique 12-digit code based on the CSI MasterFormat classification.

```
                    ┌── MasterFormat Division (03)
                    │
                    ├── MasterFormat Level 2 (03 30 00)
                    │
                    ├── MasterFormat Level 3
                    │
  03  30  53.40    3920
        │
        └── MasterFormat Level 4

        └── RSMeans 12-Digit Line Number
```

3 Description
(Concrete In Place, etc.)

Each line item is described in detail. Sub-items and additional sizes are indented beneath the appropriate line items. The first line or two after the main item (in boldface) may contain descriptive information that pertains to all line items beneath this boldface listing.

Items which include the symbol **CN** are updated in The *RSMeans Quarterly Update Service* online. To obtain access to this service contact RSMeans customer service at 1-800-334-3509.

4 Reference Number
Information

R033053 -50 You'll see reference numbers shown in shaded boxes at the beginning of some sections. These refer to related items in the Reference Section, visually identified by a vertical gray bar on the page edges.

The relation may be an estimating procedure that should be read before estimating, or technical information.

The "R" designates the Reference Section. The numbers refer to the MasterFormat 2004 classification system.

It is strongly recommended that you review all reference numbers that appear within the section in which you are working.

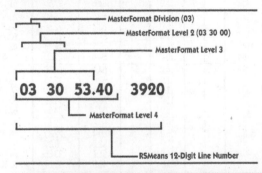

03 30 Cast-In-Place Concrete

03 30 53 – Miscellaneous Cast-In-Place Concrete

03 30 53.40 Concrete In Place		Crew	Daily Output	Labor-Hours	Unit	Material	Labor	2007 Bare Costs Equipment	Total	Total Incl O&P
3590	10' x 10' x 12" thick	C-14H	5	9.600	Ea.	565	350	4.32	919.32	1,175
3800	Footings, spread under 1 C.Y.	C-14C	38	.942	C.Y.	195	103	.56	298.56	370
3850	Over 5 C.Y.		81.04	1.382		266	48.50	.26	314.76	370
3900	Footings, strip, 18" x 9", unreinforced		40	2.800		126	98.50	.53	225.03	293
3920	18" x 9", reinforced		35	3.200		147	112	.61	259.61	340
3925	20" x 10", unreinforced		45	2.489		122	87.50	.47	209.97	273
3930	20" x 10", reinforced		40	2.800		140	98.50	.53	239.03	310
3935	24" x 12", unreinforced		55	2.036		120	71.50	.39	191.89	244
3940	24" x 12", reinforced		48	2.333		138	82	.44	220.44	280
3945	36" x 12", unreinforced		70	1.600		117	56	.30	173.30	216
3950	36" x 12", reinforced		60	1.867		133	65.50	.35	198.85	249

Figure 7.2

Credit: *Means Building Construction Cost Data 2007*

Trade Abbreviations

Abbr.	Trade	Base Rate Incl. Fringes		Workers' Comp. Ins.	Average Fixed Overhead	Overhead	Profit	Total Overhead & Profit		Rate with O & P	
		Hourly	Daily					%	Amount	Hourly	Daily
Skwk	Skilled Workers Average (35 trades)	$38.00	$304.00	16.3%	16.3%	13.0%	10%	55.6%	$21.15	$59.15	$473.20
	Helpers Average (5 trades)	27.35	218.80	17.8		11.0		55.1	15.05	42.40	339.20
	Foreman Average, Inside ($.50 over trade) Foreman	38.50	308.00	16.3		13.0		55.6	21.40	59.90	479.20
	Foreman Average, Outside ($2.00 over trade)	40.00	320.00	16.3		13.0		55.6	22.25	62.25	498.00
Clab	Common Building Laborers	28.75	230.00	18.4		11.0		55.7	16.00	44.75	358.00
Asbe	Asbestos/Insulation Workers/Pipe Coverers	41.50	332.00	15.8		16.0		58.1	24.10	65.60	524.80
Boil	Boilermakers	47.15	377.20	12.7		16.0		55.0	25.95	73.10	584.80
Bric	Bricklayers	38.05	304.40	14.9		11.0		52.2	19.85	57.90	463.20
Brhe	Bricklayer Helpers	28.65	229.20	14.9		11.0		52.2	14.95	43.60	348.80
Carp	Carpenters	36.70	293.60	18.4		11.0		55.7	20.45	57.15	457.20
Cefi	Cement Finishers	35.55	284.40	9.6		11.0		46.9	16.65	52.20	417.60
Elec	Electricians	43.90	351.20	6.6		16.0		48.9	21.45	65.35	522.80
Elev	Elevator Constructors	53.40	427.20	6.9		16.0		49.2	26.25	79.65	637.20
Eqhv	Equipment Operators, Crane or Shovel	39.80	318.40	10.4		14.0		50.7	20.20	60.00	480.00
Eqmd	Equipment Operators, Medium Equipment	38.40	307.20	10.4		14.0		50.7	19.45	57.85	462.80
Eqlt	Equipment Operators, Light Equipment	36.85	294.80	10.4		14.0		50.7	18.70	55.55	444.40
Eqol	Equipment Operators, Oilers	33.90	271.20	10.4		14.0		50.7	17.20	51.10	408.80
Eqmm	Equipment Operators, Master Mechanics	40.05	320.40	10.4		14.0		50.7	20.30	60.35	482.80
Glaz	Glaziers	36.05	288.40	14.1		11.0		51.4	18.55	54.60	436.80
Lath	Lathers	33.70	269.60	11.6		11.0		48.9	16.50	50.20	401.60
Marb	Marble Setters	36.20	289.60	14.9		11.0		52.2	18.90	55.10	440.80
Mill	Millwrights	38.20	305.60	10.2		11.0		47.5	18.15	56.35	450.80
Mstz	Mosaic & Terrazzo Workers	35.15	281.20	9.5		11.0		46.8	16.45	51.60	412.80
Pord	Painters, Ordinary	32.70	261.60	13.2		11.0		50.5	16.50	49.20	393.60
Psst	Painters, Structural Steel	33.50	268.00	45.3		11.0		82.6	27.65	61.15	489.20
Pape	Paper Hangers	32.90	263.20	13.2		11.0		50.5	16.60	49.50	396.00
Pile	Pile Drivers	35.90	287.20	21.4		16.0		63.7	22.85	58.75	470.00
Plas	Plasterers	33.55	268.40	14.0		11.0		51.3	17.20	50.75	406.00
Plah	Plasterer Helpers	28.75	230.00	14.0		11.0		51.3	14.75	43.50	348.00
Plum	Plumbers	44.80	358.40	8.1		16.0		50.4	22.60	67.40	539.20
Rodm	Rodmen (Reinforcing)	41.30	330.40	23.8		14.0		64.1	26.45	67.75	542.00
Rofc	Roofers, Composition	31.80	254.40	32.3		11.0		69.6	22.15	53.95	431.60
Rots	Roofers, Tile & Slate	31.95	255.60	32.3		11.0		69.6	22.25	54.20	433.60
Rohe	Roofers, Helpers (Composition)	23.35	186.80	32.3		11.0		69.6	16.25	39.60	316.80
Shee	Sheet Metal Workers	43.55	348.40	11.9		16.0		54.2	23.60	67.15	537.20
Spri	Sprinkler Installers	44.30	354.40	8.3		16.0		50.6	22.40	66.70	533.60
Stpi	Steamfitters or Pipefitters	45.20	361.60	8.1		16.0		50.4	22.80	68.00	544.00
Ston	Stone Masons	38.30	306.40	14.9		11.0		52.2	20.00	58.30	466.40
Sswk	Structural Steel Workers	41.35	330.80	40.8		14.0		81.1	33.55	74.90	599.20
Tilf	Tile Layers	35.50	284.00	9.5		11.0		46.8	16.60	52.10	416.80
Tilh	Tile Layers Helpers	27.25	218.00	9.5		11.0		46.8	12.75	40.00	320.00
Trlt	Truck Drivers, Light	28.55	228.40	17.1		11.0		54.4	15.55	44.10	352.80
Trhv	Truck Drivers, Heavy	29.55	236.40	17.1		11.0		54.4	16.10	45.65	365.20
Sswl	Welders, Structural Steel	41.35	330.80	40.8		14.0		81.1	33.55	74.90	599.20
Wrck	*Wrecking	28.75	230.00	39.1		11.0		76.4	21.95	50.70	405.60

*Not included in averages

Building Construction Cost Data

Figure 7.3

Credit: *Means Building Construction Cost Data 2007*

Crews

Crew No.	Bare Costs Hr.	Daily	Incl. Subs O & P Hr.	Daily	Cost Per Labor-Hour Bare Costs	Incl. O&P
Crew B-25B						
1 Labor Foreman	$30.75	$246.00	$47.90	$383.20	$32.13	$49.38
7 Laborers	28.75	1610.00	44.75	2506.00		
4 Equip. Oper. (medium)	38.40	1228.80	57.85	1851.20		
1 Asphalt Paver, 130 H.P.		1750.00		1925.00		
2 Tandem Rollers, 10 Ton		422.00		464.20		
1 Roller, Pneum. Whl, 12 Ton		295.60		325.16	25.70	28.27
96 L.H., Daily Totals		$5552.40		$7454.76	$57.84	$77.65

Crew No.	Bare Costs Hr.	Daily	Incl. Subs O & P Hr.	Daily	Cost Per Labor-Hour Bare Costs	Incl. O&P
Crew B-25C						
1 Labor Foreman	$30.75	$246.00	$47.90	$383.20	$32.30	$49.64
3 Laborers	28.75	690.00	44.75	1074.00		
2 Equip. Oper. (medium)	38.40	614.40	57.85	925.60		
1 Asphalt Paver, 130 H.P.		1750.00		1925.00		
1 Tandem Roller, 10 Ton		211.00		232.10	40.85	44.94
48 L.H., Daily Totals		$3511.40		$4539.90	$73.15	$94.58

Crew No.	Bare Costs Hr.	Daily	Incl. Subs O & P Hr.	Daily	Cost Per Labor-Hour Bare Costs	Incl. O&P
Crew B-26						
1 Labor Foreman (outside)	$30.75	$246.00	$47.90	$383.20	$32.45	$50.19
6 Laborers	28.75	1380.00	44.75	2148.00		
2 Equip. Oper. (med.)	38.40	614.40	57.85	925.60		
1 Rodman (reinf.)	41.30	330.40	67.75	542.00		
1 Cement Finisher	35.55	284.40	52.20	417.60		
1 Grader, 30,000 Lbs.		505.40		555.94		
1 Paving Mach. & Equip.		2324.00		2556.40	32.15	35.37
88 L.H., Daily Totals		$5684.60		$7528.74	$64.60	$85.55

Crew No.	Bare Costs Hr.	Daily	Incl. Subs O & P Hr.	Daily	Cost Per Labor-Hour Bare Costs	Incl. O&P
Crew B-26A						
1 Labor Foreman (outside)	$30.75	$246.00	$47.90	$383.20	$32.45	$50.19
6 Laborers	28.75	1380.00	44.75	2148.00		
2 Equip. Oper. (med.)	38.40	614.40	57.85	925.60		
1 Rodman (reinf.)	41.30	330.40	67.75	542.00		
1 Cement Finisher	35.55	284.40	52.20	417.60		
1 Grader, 30,000 Lbs.		505.40		555.94		
1 Paving Mach. & Equip.		2324.00		2556.40		
1 Concrete Saw		118.40		130.24	33.50	36.85
88 L.H., Daily Totals		$5803.00		$7658.98	$65.94	$87.03

Crew No.	Bare Costs Hr.	Daily	Incl. Subs O & P Hr.	Daily	Cost Per Labor-Hour Bare Costs	Incl. O&P
Crew B-26B						
1 Labor Foreman (outside)	$30.75	$246.00	$47.90	$383.20	$32.94	$50.83
6 Laborers	28.75	1380.00	44.75	2148.00		
3 Equip. Oper. (med.)	38.40	921.60	57.85	1388.40		
1 Rodman (reinf.)	41.30	330.40	67.75	542.00		
1 Cement Finisher	35.55	284.40	52.20	417.60		
1 Grader, 30,000 Lbs.		505.40		555.94		
1 Paving Mach. & Equip.		2324.00		2556.40		
1 Concrete Pump, 110' Boom		938.60		1032.46	39.25	43.17
96 L.H., Daily Totals		$6930.40		$9024.00	$72.19	$94.00

Crew No.	Bare Costs Hr.	Daily	Incl. Subs O & P Hr.	Daily	Cost Per Labor-Hour Bare Costs	Incl. O&P
Crew B-27						
1 Labor Foreman (outside)	$30.75	$246.00	$47.90	$383.20	$29.25	$45.54
3 Laborers	28.75	690.00	44.75	1074.00		
1 Berm Machine		250.80		275.88	7.84	8.62
32 L.H., Daily Totals		$1186.80		$1733.08	$37.09	$54.16

Crew No.	Bare Costs Hr.	Daily	Incl. Subs O & P Hr.	Daily	Cost Per Labor-Hour Bare Costs	Incl. O&P
Crew B-28						
2 Carpenters	$36.70	$587.20	$57.15	$914.40	$34.05	$53.02
1 Laborer	28.75	230.00	44.75	358.00		
24 L.H., Daily Totals		$817.20		$1272.40	$34.05	$53.02

Crew No.	Bare Costs Hr.	Daily	Incl. Subs O & P Hr.	Daily	Cost Per Labor-Hour Bare Costs	Incl. O&P
Crew B-29						
1 Labor Foreman (outside)	$30.75	$246.00	$47.90	$383.20	$31.35	$48.29
4 Laborers	28.75	920.00	44.75	1432.00		
1 Equip. Oper. (crane)	39.80	318.40	60.00	480.00		
1 Equip. Oper. Oiler	33.90	271.20	51.10	408.80		
1 Gradall, 5/8 C.Y.		905.80		996.38	16.18	17.79
56 L.H., Daily Totals		$2661.40		$3700.38	$47.52	$66.08

Crew No.	Bare Costs Hr.	Daily	Incl. Subs O & P Hr.	Daily	Cost Per Labor-Hour Bare Costs	Incl. O&P
Crew B-30						
1 Equip. Oper. (med.)	$38.40	$307.20	$57.85	$462.80	$32.50	$49.72
2 Truck Drivers (heavy)	29.55	472.80	45.65	730.40		
1 Hyd. Excavator, 1.5 C.Y.		775.60		853.16		
2 Dump Trucks, 16 Ton, 12 C.Y.		1060.00		1166.00	76.48	84.13
24 L.H., Daily Totals		$2615.60		$3212.36	$108.98	$133.85

Crew No.	Bare Costs Hr.	Daily	Incl. Subs O & P Hr.	Daily	Cost Per Labor-Hour Bare Costs	Incl. O&P
Crew B-31						
1 Labor Foreman (outside)	$30.75	$246.00	$47.90	$383.20	$30.74	$47.86
3 Laborers	28.75	690.00	44.75	1074.00		
1 Carpenter	36.70	293.60	57.15	457.20		
1 Air Compressor, 250 C.F.M.		151.40		166.54		
1 Sheeting Driver		5.75		6.33		
2 -50' Air Hoses, 1.5"		11.60		12.76	4.22	4.64
40 L.H., Daily Totals		$1398.35		$2100.03	$34.96	$52.50

Crew No.	Bare Costs Hr.	Daily	Incl. Subs O & P Hr.	Daily	Cost Per Labor-Hour Bare Costs	Incl. O&P
Crew B-32						
1 Laborer	$28.75	$230.00	$44.75	$358.00	$35.99	$54.58
3 Equip. Oper. (med.)	38.40	921.60	57.85	1388.40		
1 Grader, 30,000 Lbs.		505.40		555.94		
1 Tandem Roller, 10 Ton		211.00		232.10		
1 Dozer, 200 H.P.		988.40		1087.24	53.27	58.60
32 L.H., Daily Totals		$2856.40		$3621.68	$89.26	$113.18

Crew No.	Bare Costs Hr.	Daily	Incl. Subs O & P Hr.	Daily	Cost Per Labor-Hour Bare Costs	Incl. O&P
Crew B-32A						
1 Laborer	$28.75	$230.00	$44.75	$358.00	$35.18	$53.48
2 Equip. Oper. (medium)	38.40	614.40	57.85	925.60		
1 Grader, 30,000 Lbs.		505.40		555.94		
1 Roller, Vibratory, 25 Ton		565.60		622.16	44.63	49.09
24 L.H., Daily Totals		$1915.40		$2461.70	$79.81	$102.57

Crew No.	Bare Costs Hr.	Daily	Incl. Subs O & P Hr.	Daily	Cost Per Labor-Hour Bare Costs	Incl. O&P
Crew B-32B						
1 Laborer	$28.75	$230.00	$44.75	$358.00	$35.18	$53.48
2 Equip. Oper. (medium)	38.40	614.40	57.85	925.60		
1 Dozer, 200 H.P.		988.40		1087.24		
1 Roller, Vibratory, 25 Ton		565.60		622.16	64.75	71.22
24 L.H., Daily Totals		$2398.40		$2993.00	$99.93	$124.71

Crew No.	Bare Costs Hr.	Daily	Incl. Subs O & P Hr.	Daily	Cost Per Labor-Hour Bare Costs	Incl. O&P
Crew B-32C						
1 Labor Foreman	$30.75	$246.00	$47.90	$383.20	$33.91	$51.83
2 Laborers	28.75	460.00	44.75	716.00		
3 Equip. Oper. (medium)	38.40	921.60	57.85	1388.40		
1 Grader, 30,000 Lbs.		505.40		555.94		
1 Tandem Roller, 10 Ton		211.00		232.10		
1 Dozer, 200 H.P.		988.40		1087.24	35.52	39.07
48 L.H., Daily Totals		$3332.40		$4362.88	$69.42	$90.89

Crew No.	Bare Costs Hr.	Daily	Incl. Subs O & P Hr.	Daily	Cost Per Labor-Hour Bare Costs	Incl. O&P
Crew B-33A						
1 Equip. Oper. (med.)	$38.40	$307.20	$57.85	$462.80	$35.64	$54.11
.5 Laborer	28.75	115.00	44.75	179.00		
.25 Equip. Oper. (med.)	38.40	76.80	57.85	115.70		
1 Scraper, Towed, 7 C.Y.		143.20		157.52		
1.25 Dozer, 300 H.P.		1626.25		1788.88	126.39	139.03
14 L.H., Daily Totals		$2268.45		$2703.90	$162.03	$193.14

Figure 7.4

Credit: *Means Building Construction Cost Data 2007*

Repair & Remodeling

Cost figures in *Means Building Construction Cost Data* are based on new construction utilizing the most cost-effective combination of labor, equipment, and materials. The work is scheduled in the proper sequence to allow the various trades to accomplish their tasks in an efficient manner.

There are many factors unique to repair and remodeling that can affect project costs. The economy of scale associated with new construction often has no influence on the cost of repair and remodeling. Small quantities of components may have to be custom-fabricated at great expense. Work schedule coordination between trades frequently becomes difficult; work area restrictions can lead to subcontractor quotations with start-up and shutdown costs in excess of the cost of the actual work involved. Some of the more prominent factors affecting repair and remodeling projects are listed below:

1. A large amount of cutting and patching may be required to match the existing construction. It is often more economical to remove entire walls rather than create many new door and window openings. This sort of tradeoff has to be carefully analyzed. Matching "existing conditions" may be impossible, because certain materials may no longer be manufactured, and substitutions can be expensive. Piping and ductwork runs may not be as simple as they are in the case of new construction. Wiring may have to be snaked through existing walls and floors.

2. Dust and noise protection of adjoining non-construction areas can involve a substantial number of special precautions and may alter normal construction methods.

3. The use of certain equipment may be curtailed as a result of the physical limitations of the project; workers may be forced to use small equipment or hand tools.

4. Material handling becomes more costly due to the confines of an enclosed building. For multi-story construction, low capacity elevators and stairwells may be the only access to the upper floors.

5. Both existing and completed finish work will need to be protected in order to prevent damage during ensuing work.

6. Work may have to be done on other than normal shifts—around an existing production facility that has to stay in operation throughout the repair and remodeling. Costs for overtime work may be incurred as a result.

7. There may be an increased requirement for shoring and bracing to support the building while structural changes are being made, or to allow for the temporary storage of construction materials on above-grade floors.

These factors and their consequences can significantly increase the costs for repair and remodeling work as compared with new construction. RSMeans has developed a method to quantify these factors by adding percentages to the costs of work that is affected. These suggested percentages are shown in Figure 7.11 as minimums and maximums. The estimator must use sound judgment and experience when applying these factors. The effects of each of these factors should be considered in the planning, bidding, and construction phases in order to minimize the potential increased costs associated with repair and remodeling projects.

There are other considerations to be anticipated in estimating for repair and remodeling. Weather protection may be required for existing structures and during the installation of new windows or heating systems. Pedestrian protection is often required in urban areas. On small projects and because of local conditions, it may be necessary to pay a tradesman for a minimum of four hours for a task that actually requires less time. Unit prices should be used with caution in situations when these kinds of minimum charges may be incurred.

All of the above factors can be anticipated and the basic costs developed before a repair and remodeling project begins. It is the hidden problems— the unknowns—that pose the greatest challenge to the estimator and cause the most anxiety. These problems are often discovered during demolition and may be impossible to anticipate. Projects may be delayed due to these unexpected conditions, and these delays ultimately increase construction costs. Other parts of the project, and thus their cost, are also affected. Only experience, good judgment, and a thorough knowledge of the existing structure can help to reduce the number of unknowns and their potential effects.

City Cost Indexes

The unit prices in *Means Building Construction Cost Data* are national averages. When they are to be applied to a particular location, these prices must be adjusted to local conditions. RSMeans has developed the City Cost Indexes for just that purpose. Following the crews table in the back of *Means Building Construction Cost Data* are tables of indexes for 731 U.S. and Canadian cities based on a 30 major city average of 100. The figures are broken down into material and installation for the 49 MasterFormat divisions, as shown in Figure 7.12. Please note that for each city there is a weighted average for the material, installation, and total indexes. This average is based on the relative contribution of each division to the construction process as a whole. The information in Figure 7.13 does not represent any one building type but, instead, all building types as a whole. The figures may be used as a general guide to determine how much time should be spent on each portion of an estimate. When doing an estimate, more time should be spent on the divisions that have a higher percent contribution to the project. Caution should be exercised when using set percentages for projects that have unusually high or low division contributions.

01 21 Allowances

01 21 53 – Factors Allowance

01 21 53.50 Factors

0010	**FACTORS** Cost adjustments										
0100	Add to construction costs for particular job requirements										
0500	Cut & patch to match existing construction, add, minimum				Costs		2%	3%			
0550	Maximum						5%	9%			
0800	Dust protection, add, minimum						1%	2%			
0850	Maximum						4%	11%			
1100	Equipment usage curtailment, add, minimum						1%	1%			
1150	Maximum						3%	10%			
1400	Material handling & storage limitation, add, minimum						1%	1%			
1450	Maximum						6%	7%			
1700	Protection of existing work, add, minimum						2%	2%			
1750	Maximum						5%	7%			
2000	Shift work requirements, add, minimum							5%			
2050	Maximum							30%			
2300	Temporary shoring and bracing, add, minimum						2%	5%			
2350	Maximum						5%	12%			
2400	Work inside prisons and high security areas, add, minimum							30%			
2450	Maximum							50%			

Figure 7.11

Credit: *Means Electrical Cost Data 2007*

City Cost Indexes

Figure 7.12a

ALASKA / ARIZONA

DIVISION		KETCHIKAN 999 MAT.	INST.	TOTAL	CHAMBERS 865 MAT.	INST.	TOTAL	FLAGSTAFF 860 MAT.	INST.	TOTAL	GLOBE 855 MAT.	INST.	TOTAL	KINGMAN 864 MAT.	INST.	TOTAL	MESA/TEMPE 852 MAT.	INST.	TOTAL
01543	CONTRACTOR EQUIPMENT	.0	118.7	118.7	.0	94.6	94.6	.0	94.6	94.6	.0	98.2	98.2	.0	94.6	94.6	.0	98.2	98.2
0241, 31 - 34	SITE & INFRASTRUCTURE, DEMOLITION	196.0	133.7	151.3	64.6	98.9	89.2	81.9	99.9	94.8	95.3	102.8	100.7	64.5	100.4	90.2	86.2	103.2	98.4
0310	Concrete Forming & Accessories	126.2	115.8	117.3	96.8	53.1	59.3	102.6	66.2	71.3	96.6	52.9	59.1	94.6	64.8	69.0	99.7	61.1	66.5
0310	Concrete Reinforcing	114.0	106.8	110.5	105.3	72.4	89.2	105.1	73.2	89.5	109.9	72.3	91.5	105.4	73.1	89.7	110.6	73.2	92.3
0330	Cast-in-Place Concrete	302.2	115.9	232.1	95.9	60.8	82.7	96.0	75.1	88.2	104.3	61.1	88.1	95.6	61.8	82.9	105.2	67.9	91.1
03	CONCRETE	218.7	113.5	169.1	103.2	59.5	82.6	123.2	70.5	98.3	120.7	59.6	91.8	102.8	65.3	85.1	112.0	65.8	90.2
04	MASONRY	227.3	120.7	163.5	98.6	48.5	68.6	98.3	63.9	77.7	110.4	48.5	73.4	98.6	60.7	75.9	110.5	50.3	74.5
05	METALS	125.6	102.1	118.5	97.2	65.0	87.5	97.7	69.2	89.1	100.3	65.4	89.8	97.8	69.1	89.1	100.6	69.9	91.3
06	WOOD, PLASTICS & COMPOSITES	109.6	114.2	112.0	101.0	51.3	74.7	107.3	66.1	85.5	96.0	51.4	72.5	97.1	66.1	80.7	99.6	66.3	82.0
07	THERMAL & MOISTURE PROTECTION	174.3	114.9	150.4	96.5	59.2	81.5	98.2	69.0	86.4	102.1	56.6	83.8	96.5	64.7	83.7	101.4	61.4	85.3
08	OPENINGS	122.9	110.6	119.7	101.7	55.2	89.7	101.8	66.5	92.7	98.5	55.3	87.3	101.9	66.5	92.7	98.6	66.5	90.3
0920	Plaster & Gypsum Board	136.0	114.4	123.1	90.6	50.0	66.4	93.0	65.2	76.5	92.8	50.0	67.3	87.2	65.2	74.1	94.4	65.2	77.0
0950, 0980	Ceilings & Acoustic Treatment	125.9	114.4	119.0	104.2	50.0	71.6	105.1	65.2	81.1	95.1	50.0	68.0	105.1	65.2	81.1	95.1	65.2	77.1
0960	Flooring	164.0	125.2	153.7	91.7	47.6	80.0	94.3	47.8	81.9	91.8	47.6	80.1	90.7	64.6	83.8	93.3	59.6	84.3
0990	Wall Finishes & Painting/Coating	161.6	115.4	134.0	91.3	45.6	64.0	91.3	56.0	70.2	95.2	45.6	65.6	91.3	56.0	70.2	95.2	56.0	71.8
09	FINISHES	154.4	117.9	135.6	94.2	50.4	71.6	97.1	61.1	78.5	95.2	50.5	72.1	93.6	63.2	77.9	95.0	59.7	76.8
COVERS	DIVS. 10 - 14, 25, 28, 41, 43, 44	100.0	110.8	102.2	100.0	79.9	95.9	100.0	82.9	96.5	100.0	80.3	95.9	100.0	81.6	96.2	100.0	78.1	95.5
22, 23	PLUMBING & HVAC	98.9	99.6	99.2	97.3	72.3	87.1	100.1	77.1	90.7	95.9	65.9	83.7	97.3	75.4	88.4	100.2	66.8	86.6
26, 27, 3370	ELECTRICAL, COMMUNICATIONS & UTIL.	148.4	112.8	131.4	99.5	74.6	87.6	98.6	61.4	80.7	94.6	66.3	81.0	99.5	43.2	72.5	91.6	61.4	77.1
MF2004	WEIGHTED AVERAGE	141.6	112.4	129.0	97.8	66.2	84.2	101.5	71.3	88.4	100.8	64.1	84.9	97.8	67.5	84.7	100.2	66.9	85.8

ARIZONA / ARKANSAS

DIVISION		PHOENIX 850,853 MAT.	INST.	TOTAL	PRESCOTT 863 MAT.	INST.	TOTAL	SHOW LOW 859 MAT.	INST.	TOTAL	TUCSON 856 - 857 MAT.	INST.	TOTAL	BATESVILLE 725 MAT.	INST.	TOTAL	CAMDEN 717 MAT.	INST.	TOTAL
01543	CONTRACTOR EQUIPMENT	.0	98.8	98.8	.0	94.6	94.6	.0	98.2	98.2	.0	98.2	98.2	.0	86.5	86.5	.0	86.5	86.5
0241, 31 - 34	SITE & INFRASTRUCTURE, DEMOLITION	86.6	104.0	99.1	70.5	98.9	90.9	97.3	102.8	101.3	82.7	103.8	97.8	73.8	84.0	81.1	74.7	83.6	81.0
0310	Concrete Forming & Accessories	100.8	70.2	74.6	98.3	52.9	59.3	104.1	53.3	60.5	100.2	69.8	74.1	83.6	55.9	59.8	83.3	36.8	43.4
0310	Concrete Reinforcing	108.9	73.3	91.5	105.1	72.7	89.3	110.6	72.5	92.0	91.3	73.2	82.4	95.8	76.3	86.3	97.1	51.6	74.9
0330	Cast-in-Place Concrete	105.3	75.7	94.1	95.9	60.7	82.7	104.3	61.3	88.1	108.3	75.5	95.9	79.2	52.3	69.1	81.2	41.7	66.3
03	CONCRETE	111.6	72.6	93.2	108.6	59.4	85.4	123.1	59.8	93.2	110.2	72.3	92.3	80.1	59.0	70.1	81.7	42.2	63.0
04	MASONRY	98.1	64.9	78.2	98.6	54.7	72.3	110.4	50.3	74.5	95.9	63.9	76.8	99.3	50.8	70.3	108.0	38.4	66.4
05	METALS	102.1	71.0	92.7	97.7	65.2	87.9	100.1	66.0	89.8	101.3	69.8	91.8	96.3	70.2	88.4	96.3	60.2	85.4
06	WOOD, PLASTICS & COMPOSITES	100.6	71.3	85.1	102.5	51.3	75.4	104.4	51.4	76.4	99.9	71.3	84.8	84.5	57.2	70.0	84.2	37.0	59.2
07	THERMAL & MOISTURE PROTECTION	101.3	69.5	88.5	97.0	58.4	81.5	102.3	59.5	85.1	102.9	66.1	88.1	98.3	53.3	80.2	98.2	38.3	74.1
08	OPENINGS	99.7	69.3	91.8	101.9	55.2	89.8	97.5	55.3	86.6	94.8	69.3	88.2	96.2	55.3	85.6	92.3	43.0	79.6
0920	Plaster & Gypsum Board	96.2	70.4	80.9	90.7	50.0	66.5	96.0	50.0	68.6	99.1	70.4	82.0	84.5	56.7	67.9	84.5	35.9	55.5
0950, 0980	Ceilings & Acoustic Treatment	101.9	70.4	82.9	103.4	50.0	71.3	95.1	50.0	68.0	98.8	70.4	81.7	90.4	56.7	70.1	90.4	35.9	57.6
0960	Flooring	93.6	62.3	85.2	92.6	47.6	80.6	95.1	54.2	84.2	94.9	47.8	82.4	104.7	71.1	95.7	104.6	47.7	89.4
0990	Wall Finishes & Painting/Coating	95.2	56.5	72.1	91.3	45.6	64.0	95.2	45.6	65.6	95.4	56.0	71.8	98.3	50.9	70.0	98.3	55.5	72.7
09	FINISHES	97.1	66.9	81.5	94.7	50.4	71.8	96.9	51.6	73.5	97.1	64.1	80.1	92.7	58.3	74.9	92.7	40.6	65.8
COVERS	DIVS. 10 - 14, 25, 28, 41, 43, 44	100.0	83.9	96.7	100.0	79.8	95.9	100.0	80.3	95.9	100.0	83.9	96.7	100.0	52.8	90.3	100.0	46.9	89.1
22, 23	PLUMBING & HVAC	100.2	77.1	90.8	100.1	71.9	88.6	100.1	72.4	86.3	100.1	68.1	87.1	95.8	44.4	74.9	95.8	36.0	71.5
26, 27, 3370	ELECTRICAL, COMMUNICATIONS & UTIL.	100.0	67.0	84.2	98.3	61.3	80.6	91.9	66.3	79.6	93.6	59.3	77.2	99.1	50.9	76.0	95.9	45.4	71.7
MF2004	WEIGHTED AVERAGE	101.0	73.9	89.3	99.2	64.9	84.4	100.9	65.9	85.8	99.4	70.2	86.8	94.1	56.5	77.9	94.1	46.0	73.3

ARKANSAS

DIVISION		FAYETTEVILLE 727 MAT.	INST.	TOTAL	FORT SMITH 729 MAT.	INST.	TOTAL	HARRISON 726 MAT.	INST.	TOTAL	HOT SPRINGS 719 MAT.	INST.	TOTAL	JONESBORO 724 MAT.	INST.	TOTAL	LITTLE ROCK 720 - 722 MAT.	INST.	TOTAL
01543	CONTRACTOR EQUIPMENT	.0	86.5	86.5	.0	86.5	86.5	.0	86.5	86.5	.0	86.5	86.5	.0	108.2	108.2	.0	86.5	86.5
0241, 31 - 34	SITE & INFRASTRUCTURE, DEMOLITION	73.3	84.2	81.1	78.2	84.2	82.5	78.9	84.1	82.6	78.5	83.6	82.2	101.5	99.5	100.0	77.8	84.2	82.4
0310	Concrete Forming & Accessories	78.8	46.2	50.8	100.5	43.5	51.5	88.6	56.2	60.8	80.4	43.5	48.7	87.3	60.0	63.9	94.7	73.8	76.8
0310	Concrete Reinforcing	95.9	73.5	84.9	96.9	71.2	84.3	95.4	76.2	86.0	95.3	66.5	81.2	92.7	77.2	85.1	97.1	69.5	83.6
0330	Cast-in-Place Concrete	79.2	48.9	67.8	90.6	70.9	83.2	87.9	50.0	73.6	83.1	43.1	68.1	86.3	64.5	78.1	88.6	71.2	82.1
03	CONCRETE	79.8	53.0	67.1	87.9	58.9	74.2	86.8	58.3	73.3	84.5	48.5	67.5	84.4	66.0	75.7	86.6	72.1	79.7
04	MASONRY	89.4	47.4	64.3	95.1	60.5	74.4	99.7	48.6	69.1	80.3	31.8	51.3	90.8	52.2	67.7	93.3	60.5	73.7
05	METALS	96.3	69.4	88.1	98.3	70.3	89.9	97.3	70.2	89.1	96.3	65.9	87.1	92.7	85.4	90.5	94.3	70.3	87.0
06	WOOD, PLASTICS & COMPOSITES	79.2	44.4	60.8	103.5	38.8	69.3	90.5	57.2	72.9	80.6	44.7	61.6	88.1	60.8	73.7	100.2	79.0	89.0
07	THERMAL & MOISTURE PROTECTION	99.5	53.0	80.8	99.9	55.8	82.1	98.7	51.7	79.8	98.4	42.7	76.0	102.6	59.9	85.4	98.7	60.0	83.1
08	OPENINGS	96.2	52.6	84.9	97.0	46.9	84.0	97.0	56.1	86.4	92.3	48.9	81.1	98.8	65.0	90.0	97.0	69.7	89.9
0920	Plaster & Gypsum Board	81.9	43.5	59.1	88.6	37.7	58.3	87.7	56.7	69.2	82.9	43.8	59.6	100.5	59.9	76.3	88.6	79.1	82.9
0950, 0980	Ceilings & Acoustic Treatment	90.4	43.5	62.2	95.5	37.7	60.7	94.6	56.7	71.8	90.4	43.8	62.3	94.3	59.9	73.6	95.5	79.1	85.6
0960	Flooring	101.2	71.1	93.1	113.5	71.8	102.4	107.8	71.1	98.0	103.0	71.1	94.0	75.8	64.3	72.8	115.0	71.8	103.5
0990	Wall Finishes & Painting/Coating	98.3	35.5	60.8	98.3	66.5	79.3	98.3	50.9	70.0	98.3	44.1	65.9	86.9	59.8	70.7	98.3	68.3	80.4
09	FINISHES	91.3	48.7	69.3	97.4	50.2	73.0	95.4	58.3	76.3	92.3	48.3	69.6	90.3	60.5	74.9	97.8	74.0	85.5
COVERS	DIVS. 10 - 14, 25, 28, 41, 43, 44	100.0	64.4	92.7	100.0	71.3	94.1	100.0	66.5	93.1	100.0	48.5	89.4	100.0	60.6	91.9	100.0	76.2	95.1
22, 23	PLUMBING & HVAC	95.9	49.2	76.8	100.1	52.2	80.6	95.8	51.3	77.7	95.8	41.3	73.6	100.3	52.2	80.7	100.1	72.2	88.7
26, 27, 3370	ELECTRICAL, COMMUNICATIONS & UTIL.	93.5	63.8	79.3	96.7	80.5	88.9	97.8	46.2	73.1	97.9	47.8	73.9	105.9	55.8	81.8	96.8	81.8	89.6
MF2004	WEIGHTED AVERAGE	92.8	57.0	77.3	96.7	62.3	81.9	95.5	57.4	79.0	93.2	49.6	74.4	96.4	63.8	82.3	95.8	73.2	86.0

Figure 7.12a

CALIFORNIA

DIVISION		PALM SPRINGS 922			PALO ALTO 943			PASADENA 910 - 912			REDDING 960			RICHMOND 948			RIVERSIDE 925		
		MAT.	INST.	TOTAL	MAT.	INST.	TOTAL	MAT.	INST.	TOTAL	MAT.	INST.	TOTAL	MAT.	INST.	TOTAL	MAT.	INST.	TOTAL
01543	CONTRACTOR EQUIPMENT	.0	100.9	100.9	.0	103.0	103.0	.0	98.1	98.1	.0	99.4	99.4	.0	103.0	103.0	.0	100.9	100.9
0241, 31 - 34	SITE & INFRASTRUCTURE, DEMOLITION	94.9	106.9	103.5	138.9	103.8	113.8	100.7	109.5	107.0	115.9	105.4	108.4	152.2	103.8	117.5	102.3	106.9	105.6
0310	Concrete Forming & Accessories	101.5	119.0	116.5	104.9	139.1	134.3	103.5	118.8	116.6	103.9	125.2	122.2	123.8	139.1	137.0	105.9	118.9	117.1
0310	Concrete Reinforcing	111.9	116.6	114.2	101.7	117.4	109.4	103.9	116.7	110.2	106.4	116.7	111.4	101.7	117.4	109.4	108.7	116.6	112.5
0330	Cast-in-Place Concrete	97.2	117.4	104.8	122.5	118.0	120.8	100.1	114.5	105.5	122.7	113.9	119.4	141.0	118.0	132.3	105.8	117.4	110.1
03	CONCRETE	101.3	117.2	108.8	111.8	126.6	118.8	103.4	115.9	109.3	118.8	118.7	118.8	126.5	126.6	126.6	107.8	117.1	112.2
04	MASONRY	87.1	110.8	101.3	127.8	120.6	123.5	108.1	114.2	111.7	118.2	110.5	113.6	153.9	120.6	134.0	88.1	110.8	101.7
05	METALS	108.3	102.5	106.6	97.5	106.8	100.3	88.4	99.8	91.8	107.1	102.0	105.5	97.6	106.7	100.3	107.8	102.4	106.2
06	WOOD, PLASTICS & COMPOSITES	90.5	118.2	105.1	100.1	143.1	122.8	86.8	118.0	103.3	95.2	126.9	112.0	96.2	143.1	122.3	96.2	118.2	107.9
07	THERMAL & MOISTURE PROTECTION	102.5	113.1	106.7	110.4	128.8	117.8	99.9	111.0	104.3	105.9	112.2	108.5	111.1	127.4	117.6	102.6	113.5	107.0
08	OPENINGS	99.7	116.1	104.0	105.6	133.0	112.7	95.0	116.0	100.4	103.3	120.1	107.7	105.6	133.0	112.7	102.8	116.1	106.3
0920	Plaster & Gypsum Board	98.9	118.6	110.6	102.4	143.7	127.0	103.1	118.6	112.3	101.3	127.4	116.8	110.7	143.7	130.3	102.1	118.6	111.9
0950, 0980	Ceilings & Acoustic Treatment	112.5	118.6	116.2	121.1	143.7	134.6	96.0	118.6	109.6	126.4	127.4	127.0	121.1	143.7	134.6	117.2	118.6	118.1
0960	Flooring	119.2	107.3	116.1	116.6	121.6	117.9	99.5	107.3	101.6	116.4	117.7	116.7	125.6	124.8	125.4	121.5	107.3	117.7
0990	Wall Finishes & Painting/Coating	105.5	137.5	124.6	111.5	137.9	127.2	108.3	109.8	109.2	108.2	112.6	110.8	111.5	137.9	127.2	105.5	109.8	108.1
09	FINISHES	100.1	118.9	114.6	117.0	137.2	127.4	102.2	115.6	109.1	116.0	123.8	120.0	121.9	137.7	130.1	112.8	115.7	114.3
COVERS	DIVS. 10 - 14, 25, 28, 41, 43, 44	100.0	112.4	102.5	100.0	125.1	105.2	100.0	111.9	102.4	100.0	121.5	104.4	100.0	125.1	105.2	100.0	112.4	102.5
22, 23	PLUMBING & HVAC	95.9	106.7	100.3	96.1	135.3	112.1	95.9	106.7	100.3	100.2	108.3	103.5	96.1	132.3	110.9	100.1	107.6	103.1
26, 27, 3370	ELECTRICAL, COMMUNICATIONS & UTIL.	93.6	102.9	98.0	102.5	137.8	119.4	118.9	111.7	115.5	98.4	104.8	101.5	102.9	132.9	117.3	90.4	102.9	96.4
MF2004	WEIGHTED AVERAGE	99.8	110.1	104.3	105.2	127.4	114.8	99.9	111.0	104.7	106.4	111.8	108.7	109.3	126.1	116.6	101.9	108.9	105.3

CALIFORNIA

DIVISION		SACRAMENTO 942,956 - 958			SALINAS 939			SAN BERNARDINO 923 - 924			SAN DIEGO 919 - 921			SAN FRANCISCO 940 - 941			SAN JOSE 951		
		MAT.	INST.	TOTAL	MAT.	INST.	TOTAL	MAT.	INST.	TOTAL	MAT.	INST.	TOTAL	MAT.	INST.	TOTAL	MAT.	INST.	TOTAL
01543	CONTRACTOR EQUIPMENT	.0	102.6	102.6	.0	99.8	99.8	.0	100.9	100.9	.0	98.1	98.1	.0	108.3	108.3	.0	100.2	100.2
0241, 31 - 34	SITE & INFRASTRUCTURE, DEMOLITION	117.3	110.1	112.2	127.6	106.2	112.2	78.8	106.9	99.0	106.4	102.1	103.3	155.3	110.2	123.0	150.9	99.7	114.2
0310	Concrete Forming & Accessories	105.8	125.6	122.8	108.1	128.8	125.9	110.1	118.8	117.6	106.1	109.9	109.3	107.6	140.0	135.4	106.5	139.0	134.4
0310	Concrete Reinforcing	96.4	116.7	106.3	110.3	117.2	113.6	108.7	116.5	112.5	100.7	116.5	108.4	118.5	117.9	118.2	97.0	117.4	107.0
0330	Cast-in-Place Concrete	116.1	114.7	115.6	107.0	114.3	109.7	73.0	117.3	89.7	109.8	107.3	108.8	140.9	119.6	132.9	128.7	117.5	124.5
03	CONCRETE	111.9	119.1	115.3	118.6	120.5	119.5	80.6	117.1	97.8	109.4	109.6	109.5	128.1	127.6	127.9	117.6	126.4	121.8
04	MASONRY	129.8	110.5	118.3	111.5	117.2	114.9	95.2	110.8	104.5	99.1	109.3	105.2	163.4	126.8	141.5	149.5	120.7	132.3
05	METALS	95.5	101.5	97.3	106.9	103.3	105.8	107.8	102.2	106.1	105.0	101.5	104.0	105.5	108.8	106.5	101.6	108.3	103.6
06	WOOD, PLASTICS & COMPOSITES	97.9	127.1	113.3	100.4	130.8	116.4	100.7	118.2	110.0	100.5	107.1	104.0	103.1	143.3	124.3	102.0	142.8	123.6
07	THERMAL & MOISTURE PROTECTION	113.0	113.4	113.2	101.3	120.2	108.9	101.7	113.7	106.5	109.2	104.2	107.2	114.0	131.3	121.0	101.9	129.2	112.9
08	OPENINGS	118.9	120.2	119.2	102.1	126.4	108.4	99.8	116.1	104.0	103.3	108.7	104.7	110.2	133.1	116.1	93.9	132.9	104.0
0920	Plaster & Gypsum Board	99.5	127.4	116.1	103.8	131.4	120.2	103.6	118.6	112.6	107.8	107.1	107.4	106.2	143.7	128.5	98.3	143.7	125.3
0950, 0980	Ceilings & Acoustic Treatment	122.0	127.4	125.3	120.0	131.4	126.9	115.0	118.6	117.2	102.7	107.1	105.3	132.0	143.7	139.0	110.6	143.7	130.5
0960	Flooring	121.1	117.7	120.2	118.1	124.8	119.9	123.9	101.0	117.8	108.3	107.3	108.1	117.6	125.9	119.8	110.6	125.9	114.7
0990	Wall Finishes & Painting/Coating	108.8	112.6	111.0	109.2	137.9	126.3	105.5	109.8	108.1	105.4	109.8	108.1	111.5	149.7	134.3	110.1	137.9	126.7
09	FINISHES	116.9	123.9	120.5	116.6	130.1	123.6	111.7	114.5	113.2	107.9	109.0	108.5	121.5	139.6	130.8	111.4	137.8	125.0
COVERS	DIVS. 10 - 14, 25, 28, 41, 43, 44	100.0	122.1	104.5	100.0	122.0	104.5	100.0	108.1	101.7	100.0	111.0	102.3	100.0	125.7	105.3	100.0	124.5	105.0
22, 23	PLUMBING & HVAC	100.2	110.0	104.2	96.0	115.8	104.1	95.9	106.7	100.3	100.2	106.1	102.6	100.4	147.8	119.7	100.2	135.2	114.5
26, 27, 3370	ELECTRICAL, COMMUNICATIONS & UTIL.	97.3	104.8	100.9	89.4	120.7	104.4	93.6	105.5	99.3	100.0	97.8	99.0	102.7	151.3	126.0	102.4	137.8	119.4
MF2004	WEIGHTED AVERAGE	106.0	112.6	108.9	104.1	118.1	110.1	97.6	109.7	102.8	103.4	105.6	104.3	112.6	133.8	121.8	107.1	127.2	115.8

CALIFORNIA

DIVISION		SAN LUIS OBISPO 934			SAN MATEO 944			SAN RAFAEL 949			SANTA ANA 926 - 927			SANTA BARBARA 931			SANTA CRUZ 950		
		MAT.	INST.	TOTAL	MAT.	INST.	TOTAL	MAT.	INST.	TOTAL	MAT.	INST.	TOTAL	MAT.	INST.	TOTAL	MAT.	INST.	TOTAL
01543	CONTRACTOR EQUIPMENT	.0	99.8	99.8	.0	103.0	103.0	.0	103.1	103.1	.0	100.9	100.9	.0	99.8	99.8	.0	100.2	100.2
0241, 31 - 34	SITE & INFRASTRUCTURE, DEMOLITION	118.1	106.4	109.7	148.8	103.8	116.6	133.5	110.0	116.7	93.0	106.9	103.0	111.2	106.4	107.7	150.3	99.6	113.9
0310	Concrete Forming & Accessories	114.5	118.8	118.2	112.1	139.3	135.4	118.0	138.9	135.9	110.9	118.9	117.7	104.9	118.9	116.9	106.5	129.0	125.8
0310	Concrete Reinforcing	111.7	116.5	114.1	101.7	117.5	109.5	102.4	117.5	109.8	112.4	116.6	114.4	109.7	116.5	113.0	120.1	117.2	118.7
0330	Cast-in-Place Concrete	115.0	116.4	115.5	136.5	118.1	129.6	158.9	116.6	143.0	93.3	117.4	102.4	107.9	116.4	111.1	127.8	116.4	123.5
03	CONCRETE	117.4	116.7	117.1	122.4	126.7	124.4	146.2	125.8	136.6	98.8	117.1	107.4	109.0	116.8	112.7	120.8	121.6	121.2
04	MASONRY	113.6	110.4	111.7	153.5	123.9	135.8	125.7	123.8	124.6	83.4	111.2	100.0	83.7	111.0	100.0	154.2	117.3	132.1
05	METALS	104.7	101.9	103.8	97.4	107.2	100.4	98.6	103.0	100.0	107.9	102.4	106.2	102.7	102.0	102.5	109.6	107.4	108.9
06	WOOD, PLASTICS & COMPOSITES	104.0	118.3	111.6	108.9	143.1	127.0	109.4	142.8	127.0	102.5	118.2	110.8	95.1	118.3	107.4	102.0	130.9	117.3
07	THERMAL & MOISTURE PROTECTION	101.3	109.9	104.7	110.8	130.1	118.5	116.2	126.5	120.3	102.8	113.2	107.0	100.7	110.7	104.7	101.7	121.8	109.8
08	OPENINGS	99.8	113.4	103.4	105.6	133.0	112.7	116.0	132.8	120.4	98.9	116.1	103.4	101.8	116.2	105.5	95.1	126.4	103.4
0920	Plaster & Gypsum Board	110.9	118.6	115.5	105.9	143.7	128.4	108.8	143.7	129.5	103.9	118.6	112.7	102.7	118.6	112.2	107.4	131.4	121.7
0950, 0980	Ceilings & Acoustic Treatment	120.0	118.6	119.2	121.1	143.7	134.6	130.3	143.7	138.3	115.0	118.6	117.2	119.7	118.6	119.1	116.7	131.4	125.5
0960	Flooring	126.1	94.3	117.6	119.8	124.8	121.1	131.2	121.6	128.6	124.6	107.3	120.0	117.1	94.3	111.1	113.8	124.8	116.8
0990	Wall Finishes & Painting/Coating	108.2	103.7	105.5	111.5	137.9	127.2	107.5	137.9	125.6	105.5	109.8	108.1	108.2	103.7	105.5	109.4	137.9	126.4
09	FINISHES	119.2	112.4	115.7	119.1	137.7	128.7	122.6	137.0	130.0	113.1	115.7	114.5	114.6	112.9	113.7	115.3	130.2	123.0
COVERS	DIVS. 10 - 14, 25, 28, 41, 43, 44	100.0	121.0	104.3	100.0	125.1	105.2	100.0	124.3	105.0	100.0	112.4	102.5	100.0	112.7	102.6	100.0	122.4	104.6
22, 23	PLUMBING & HVAC	96.0	106.7	100.3	96.1	135.6	112.2	96.1	147.6	117.1	95.9	106.7	100.3	100.2	115.9	106.0	100.2	110.7	104.6
26, 27, 3370	ELECTRICAL, COMMUNICATIONS & UTIL.	88.0	99.3	93.4	102.5	141.3	121.1	96.8	114.9	105.5	93.6	105.0	99.1	87.0	107.2	96.7	101.5	120.7	110.7
MF2004	WEIGHTED AVERAGE	103.4	108.6	105.6	108.3	128.5	117.0	110.2	127.1	117.5	99.6	110.0	104.1	102.4	109.8	105.6	109.2	118.1	113.1

Figure 7.12b

Credit: *Means Building Construction Cost Data 2007*

Cost Breakdown by MasterFormat Divisions

NO.	DIVISION	%	NO.	DIVISION	%	NO.	DIVISION	%
15433	CONTRACTOR EQUIPMENT*	5.8%	0510	Structural Metal Framing	3.5%	0920	Plaster & Gypsum Board	2.1%
3120, 3130	Earth Moving & Earthwork	4.0	0520, 0530	Metal Joists & Decking	8.3	0930, 0966	Tile & Terrazzo	2.3
3160, 3170	Load Bearing Elements & Tunneling	0.1	0550	Metal Fabrications	1.0	0950, 0980	Ceilings & Acoustical Treatment	3.1
3210	Bases. Ballasts & Paving	0.3	05	METALS	12.8	0960	Flooring	2.1
3310, 3330	Utility Services & Drainage	0.2	0610	Rough Carpentry	0.8	0970, 0990	Wall Finishes, Painting / Coatings	0.9
3230	Site Improvements	0.1	0820	Finish Carpentry	0.2	09	FINISHES	10.5
3290	Planting	0.6	06	WOOD & PLASTICS	1.0	10-14	DIVISIONS 10-14, 25, 28, 41, 43, 44	5.9
0241, 31 - 34	SITE & INFRASTRUCTURE, DEMOLITION	5.3	0710	Dampproofing & Waterproofing	0.1	2210, 2230	Piping, Pumps, Plumbing Equipment	12.9
0310	Concrete Forming & Accessories	3.0	0720, 0780	Thermal, Fire & Smoke Protection	1.5	2100	Fire Supression	3.2
0320	Concrete Reinforcing	2.0				2350	Central Heating Equipment	2.2
0330	Cast-In-Place Concrete	5.3	0740, 0750	Roofing & Siding	1.1	2330, 2340	Air Conditioning & Ventilation	3.7
0340	Precast Concrete	2.1	0760	Flashing & Sheet Metal	0.4	21, 22, 23	FIRE PROTECTION, PLUMBING & HVAC	22.0
			07	THERMAL & MOISTURE PROT.	3.1	26, 27, 3370	ELECTRICAL, COMMUNICATIONS & UTILITIES	12.7
03	CONCRETE	12.4	0810, 0830	Doors & Frames	6.0		TOTAL (Div. 1-16)	100.0%
0405	Basic Masonry Materials & Methods	0.6						
0420	Unit Masonry	6.4	0840, 0880	Glazing & Curtain Walls	1.0			
0440	Sone Assemblies	0.3	08	OPENINGS	7.0			
04	MASONRY	7.3						

* Percentage for contractor equipment is spread among divisions and included above for information only

Figure 7.13

In addition to adjusting costs in *Means Building Construction Cost Data* for particular locations, the City Cost Indexes can also be used to adjust costs from one city to another. For example, the price of a particular building type is known for City A. In order to budget the costs of the same building type in City B, the following calculation can be made:

$$\frac{\text{City B Index} \times \text{City A Index}}{\text{City A Index}} = \text{City B Cost}$$

While City Cost Indexes provide a means to adjust prices for location, the Historical Cost Index, as shown in Figure 7.14, provides a means to adjust for time. Using the same principle as above, a time adjustment factor can be calculated:

$$\frac{\text{Index for Year X}}{\text{Index for Year Y}} = \text{Time Adjustment Factor}$$

This time adjustment factor can be used to determine what the budget costs would be for a particular building type in Year X, based on costs for a similar building type known from Year Y. Used in conjunction, the two indexes allow for cost adjustments from one city during a given year to another city in another year (the present or otherwise).

For example, an office building built in San Francisco in 2002 originally cost $1,570,000. How much will a similar building cost in Phoenix in 2007?

Historical Cost Indexes

The table below lists both the RSMeans Historical Cost Index based on Jan. 1, 1993 = 100 as well as the computed value of an index based on Jan. 1, 2007 costs. Since the Jan. 1, 2007 figure is estimated, space is left to write in the actual index figures as they become available through either the quarterly "RSMeans Construction Cost Indexes" or as printed in the "Engineering News-Record." To compute the actual index based on Jan. 1, 2007 = 100, divide the Historical Cost Index for a particular year by the actual Jan. 1, 2007 Construction Cost Index. Space has been left to advance the index figures as the year progresses.

Year	Historical Cost Index Jan. 1, 1993 = 100		Current Index Based on Jan. 1, 2007 = 100		Year	Historical Cost Index Jan. 1, 1993 = 100	Current Index Based on Jan. 1, 2007 = 100		Year	Historical Cost Index Jan. 1, 1993 = 100	Current Index Based on Jan. 1, 2007 = 100	
	Est.	Actual	Est.	Actual		Actual	Est.	Actual		Actual	Est.	Actual
Oct 2007					July 1992	99.4	59.8		July 1974	41.4	24.9	
July 2007					1991	96.8	58.2		1973	37.7	22.7	
April 2007					1990	94.3	56.7		1972	34.8	20.9	
Jan 2007	166.3		100.0	100.0	1989	92.1	55.4		1971	32.1	19.3	
July 2006		162.0	97.4		1988	89.9	54.0		1970	28.7	17.3	
2005		151.6	91.2		1987	87.7	52.7		1969	26.9	16.2	
2004		143.7	86.4		1986	84.2	50.6		1968	24.9	15.0	
2003		132.0	79.4		1985	82.6	49.7		1967	23.5	14.1	
2002		128.7	77.4		1984	82.0	49.3		1966	22.7	13.7	
2001		125.1	75.2		1983	80.2	48.2		1965	21.7	13.0	
2000		120.9	72.7		1982	76.1	45.8		1964	21.2	12.7	
1999		117.6	70.7		1981	70.0	42.1		1963	20.7	12.4	
1998		115.1	69.2		1980	62.9	37.8		1962	20.2	12.1	
1997		112.8	67.8		1979	57.8	34.8		1961	19.8	11.9	
1996		110.2	66.3		1978	53.5	32.2		1960	19.7	11.8	
1995		107.6	64.7		1977	49.5	29.8		1959	19.3	11.6	
1994		104.4	62.8		1976	46.9	28.2		1958	18.8	11.3	
1993		101.7	61.2		1975	44.8	26.9		1957	18.4	11.1	

Adjustments to Costs

The Historical Cost Index can be used to convert National Average building costs at a particular time to the approximate building costs for some other time.

Example:

Estimate and compare construction costs for different years in the same city.

To estimate the National Average construction cost of a building in 1970, knowing that it cost $900,000 in 2007:

INDEX in 1970 = 28.7

INDEX in 2007 = 166.3

Note: The City Cost Indexes for Canada can be used to convert U.S. National averages to local costs in Canadian dollars.

Time Adjustment using the Historical Cost Indexes:

$$\frac{\text{Index for Year A}}{\text{Index for Year B}} \times \text{Cost in Year B} = \text{Cost in Year A}$$

$$\frac{\text{INDEX 1970}}{\text{INDEX 2007}} \times \text{Cost 2007} - \text{Cost 1970}$$

$$\frac{28.7}{166.3} \times \$900,000 = .173 \times \$900,000 = \$155,700$$

The construction cost of the building in 1970 is $155,700.

Figure 7.14

Credit: Means Building Construction Cost Data 2007

Adjustment factors are developed as shown previously using data from Figures 7.12 and 7.14:

$$\frac{\text{Phoenix index}}{\text{San Francisco index}} = \frac{89.3}{121.8} = 0.73$$

$$\frac{2007 \text{ index}}{2002 \text{ index}} = \frac{166.3}{128.7} = 1.29$$

Original cost × location adjustment × time adjustment = Proposed new cost

$1,570,000 × 0.73 × 1.29 = $1,478,470

Reference Tables

Throughout the unit price pages of *Means Building Construction Cost Data*, certain line items contain reference numbers in shaded boxes (as shown in Figure 7.15), which refer the reader to expanded data and information in the back of the book. The reference table section contains over 120 pages of tables, charts, definitions, and costs, all of which clarify the unit price data. The development of many unit costs is explained and detailed in this section.

03 30 Cast-In-Place Concrete

03 30 53 – Miscellaneous Cast-In-Place Concrete

03 30 53.40 Concrete In Place		Crew	Daily Output	Labor-Hours	Unit	Material	2007 Bare Costs Labor	Equipment	Total	Total Incl O&P
6350	16' high	C-14D	91	2.198	C.Y.	149	80.50	8	237.50	298
6800	Stairs, not including safety treads, free standing, 3'-6" wide	C-14H	83	.578	LF Nose	5.70	21	.26	26.96	39
6850	Cast on ground		125	.384	"	4.52	13.95	.17	18.64	26.50
7000	Stair landings, free standing		200	.240	S.F.	4.63	8.70	.11	13.44	18.75
7050	Cast on ground		475	.101	"	3.45	3.67	.05	7.17	9.55

03 31 Structural Concrete

03 31 05 – Normal Weight Structural Concrete

03 31 05.30 Concrete, Field Mix

0010	**CONCRETE, FIELD MIX**	R033105-65								
0015	FOB forms 2250 psi					C.Y.	92.50		92.50	102
0020	3000 psi					"	96.50		96.50	106

03 31 05.35 Normal Weight Concrete, Ready Mix

0010	**NORMAL WEIGHT CONCRETE, READY MIX**	R033105-10								
0012	Includes local aggregate, sand, portland cement, and water									
0015	Excludes all additives and treatments	R033105-20								
0020	2000 psi					C.Y.	99.50		99.50	110
0100	2500 psi	R033105-30					101		101	111
0150	3000 psi	CN					104		104	114
0200	3500 psi	R033105-40					106		106	116
0300	4000 psi						108		108	119
0350	4500 psi	R033105-50					110		110	121
0400	5000 psi	CN					114		114	125
0411	6000 psi						130		130	143
0412	8000 psi						212		212	233
0413	10,000 psi						300		300	330
0414	12,000 psi						365		365	400
1000	For high early strength cement, add						10%			
1010	For structural lightweight with regular sand, add						25%			
2000	For all lightweight aggregate, add						45%			

03 31 05.70 Placing Concrete

0010	**PLACING CONCRETE**	R033105-70								
0020	Includes labor and equipment to place and vibrate									
0050	Beams, elevated, small beams, pumped	C-20	60	1.067	C.Y.		33	12.50	45.50	65
0100	With crane and bucket	C-7	45	1.600			50	25	75	105
0200	Large beams, pumped	C-20	90	.711			22	8.35	30.35	43
0250	With crane and bucket	C-7	65	1.108			35	17.25	52.25	72.50
0400	Columns, square or round, 12" thick, pumped	C-20	60	1.067			33	12.50	45.50	65
0450	With crane and bucket	C-7	40	1.800			56.50	28	84.50	118
0600	18" thick, pumped	C-20	90	.711			22	8.35	30.35	43
0650	With crane and bucket	C-7	55	1.309			41	20.50	61.50	85.50
0800	24" thick, pumped	C-20	92	.696			21.50	8.15	29.65	42
0850	With crane and bucket	C-7	70	1.029			32.50	16	48.50	67
1000	36" thick, pumped	C-20	140	.457			14.20	5.35	19.55	28
1050	With crane and bucket	C-7	100	.720			22.50	11.20	33.70	47
1400	Elevated slabs, less than 6" thick, pumped	C-20	140	.457			14.20	5.35	19.55	28
1450	With crane and bucket	C-7	95	.758			24	11.80	35.80	49.50
1500	6" to 10" thick, pumped	C-20	160	.400			12.40	4.70	17.10	24.50
1550	With crane and bucket	C-7	110	.655			20.50	10.20	30.70	42.50
1600	Slabs over 10" thick, pumped	C-20	180	.356			11.05	4.17	15.22	21.50

Figure 7.15

Chapter 8

Computerized Estimating Methods

Computers have clearly added speed, power, and accuracy to construction estimating. They make it possible to produce more estimates in less time, break a job down to a more detailed level for better cost control, manage change orders more easily, test "what if" scenarios with ease, and integrate estimating with other commonly used construction applications. The objective of this chapter is to provide an overview of unit price estimating through the use of computerized estimating methods. Figure 8.1 provides an overview of the various levels of estimating programs and their features/applications.

Automating the estimating function is an evolutionary process. There are many levels of computerized estimating, which vary in functionality, sophistication, time required to learn, and, most of all, price. Many estimators make the mistake of immediately transitioning their manual estimating into a fully integrated estimating software system without learning the basics of what a computer can do for the estimating process. Successful implementation of an estimating software system will not happen overnight. It usually takes months of training and user interaction to get a system working to its full capability.

It is recommended that computerized estimating be introduced through a multi-step process. First, learn the basics of computers and estimating through the use of industry-standard spreadsheet software programs, such as Lotus® 1-2-3, or Microsoft® Excel. Next, introduce spreadsheet add-on programs that will provide more versatility than a standard spreadsheet program. Finally, consider upgrading to a complete estimating software application. This last step usually allows estimating software to

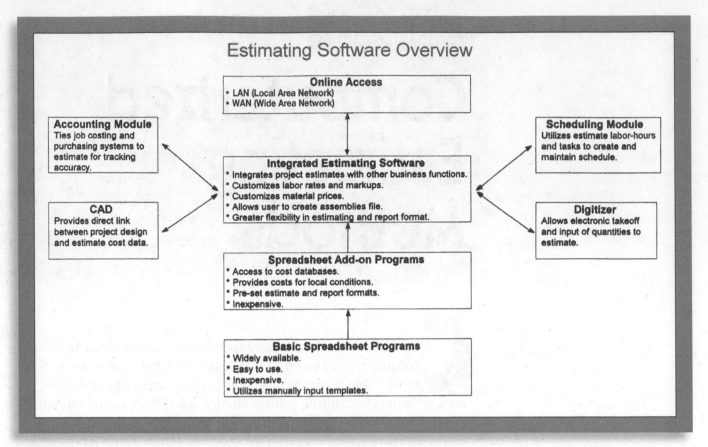

Estimating Software Overview

Online Access
* LAN (Local Area Network)
* WAN (Wide Area Network)

Accounting Module
Ties job costing and purchasing systems to estimate for tracking accuracy.

Scheduling Module
Utilizes estimate labor-hours and tasks to create and maintain schedule.

Integrated Estimating Software
* Integrates project estimates with other business functions.
* Customizes labor rates and markups.
* Customizes material prices.
* Allows user to create assemblies file.
* Greater flexibility in estimating and report format.

CAD
Provides direct link between project design and estimate cost data.

Digitizer
Allows electronic takeoff and input of quantities to estimate.

Spreadsheet Add-on Programs
* Access to cost databases.
* Provides costs for local conditions.
* Pre-set estimate and report formats.
* Inexpensive.

Basic Spreadsheet Programs
* Widely available.
* Easy to use.
* Inexpensive.
* Utilizes manually input templates.

Figure 8.1

be integrated with other related software applications, such as job cost accounting, scheduling, project management, and CAD.

The basis of any computerized estimating system is its unit price cost database. No matter how sophisticated or user-friendly construction estimating software is, its overall success depends on the completeness, functionality, and accuracy of the cost data, and the methods by which it is used. Without a fully functional database, construction estimating software is nothing more than a very expensive calculator.

One effective way to move through the different levels of computerized estimating is by using industry-standard data as the foundation of the system. Utilizing *Means CostWorks*®, for example, it is possible to seamlessly transition an estimating system from one step to the next, easily updating the cost data and integrating it into thousands of pre-built assemblies. While it is always possible to create a customized cost database from historical job costs, *Means CostWorks* will provide a foundation of costs, as well as the necessary framework to successfully estimate using a computer.

Basic Spreadsheet Programs

A computerized estimating system should perform three basic functions:

1. Calculate costs.
2. Store and manage data.
3. Generate hard copy reports.

Most industry-standard spreadsheet software programs, such as Lotus® 1-2-3 and Excel, meet these requirements. The premise of every spreadsheet program is to automate a calculation sheet. Spreadsheet programs calculate costs in vertical columns and horizontal rows using basic mathematical functions. Figure 8.2 shows a sample screen with a spreadsheet that might be used by a contractor for estimating purposes.

Everyone who has ever done construction estimating has been in the situation of rushing to finish an estimate in order to submit a bid by a certain deadline. Subcontractor quotes usually don't come in until the last minute, and it is a huge task to recalculate costs to complete the bid. Spreadsheet programs can recalculate costs in fractions of a second, which makes it fairly simple to rework the estimate prior to a bid, and allow "what if" analysis for fine-tuning.

Another major capability of spreadsheet programs is managing and sorting information. A typical line item of cost data consists of some sort of a line number, description of task, material costs, labor costs, and equipment costs. There is information on crews and productivities that needs to be maintained. Typical construction databases are in the thousands and tens of thousands of line items. Most spreadsheet programs allow information to be entered, managed, and sorted numerically or alphabetically.

When dealing with a client, estimate presentation is important. Spreadsheets make it possible to print out professional looking estimates, using different fonts, styles, and formats. Most versions also have a graphics capability. Charts and graphs can be produced based on information from an estimate spreadsheet, and a company logo can be incorporated into reports.

There are some limitations in using spreadsheets. Information within a spreadsheet must be organized using the same field layout. One problem may be locating certain line items within the database and bringing the information back into the estimate. Scrolling through a database of thousands of line items is a time-consuming and cumbersome process. Keeping such a database up-to-date also requires a major investment of time. Furthermore, it is usually not possible to apply a standardized cost database to standard spreadsheets without an add-on program (discussed in the next section). Industry-standard databases are important because they provide the ability to update information through database companies such as RSMeans.

These limitations can be overcome by features offered in the next category of computerized estimating: spreadsheet add-on programs. These are systems that provide the basic features of stand-alone (integrated)

If there is no full-time field engineer on the project, then costs may be incurred for establishing grades and setting stakes, building layout, producing as-built drawings, shop drawings, or maintenance records. With regard to surveying and layout of the building and roads, it is important to determine who is the responsible party—the owner, general contractor, or appropriate subcontractor. This responsibility should be established, if not done so in the contract documents, so that costs may be properly allocated.

Depending on the size of the project, a carpenter and/or a laborer may be assigned to the job on a full-time basis for miscellaneous work. In such cases, workers are directly responsible to the job superintendent for various tasks; the costs of this work would not be attributable to any specific division and would most appropriately be included as project overhead.

Equipment
As discussed in Chapter 4, equipment costs may be recorded as project overhead or included in each appropriate division. Some equipment, however, will be used by more than one trade; examples are personnel or material hoists and cranes. The allocation of costs in these cases should be according to company practice.

Testing
Various tests may be required by the owner, local building officials, or as specified in the contract documents. Depending on the stage of design development and the role of the contractor, soil borings and/or percolation tests may be required. Usually, this type of testing is the responsibility of the designer or engineer, paid for directly by the owner. Most testing during construction is required to verify conformance of the materials and methods to the requirements of the specifications. The most common types of testing follow.

Soil Compaction
This is usually specified as a percentage of maximum density. Strict compaction methods are required under slabs on grade and at backfill of foundation walls and footings. Testing may be required every day (or possibly every lift). Soil samples may have to be lab tested for cohesiveness, permeability, and/or water content. Watering may be required to achieve the specified compaction.

Concrete
Concrete tests may be required at two stages: slump tests during placement and compression tests after a specified curing time. Slump tests, indicating relative water and cement content, may be required for each truckload, after placement of a certain number of cubic yards, each day, or per building section. Compression tests are performed on samples placed in cylinders, usually after 7 and 28 days of curing time. The cylinders are tested by outside laboratories. If the concrete samples fail to meet design

specifications, core drilling of in-place building samples for further testing may be required.

Miscellaneous Testing

Other testing may be necessary based on the type of construction and owner or architect/engineer requirements. Core samples of asphalt paving are often required to verify specified thickness. Steel connection and weld testing may be specified for critical structural points. Masonry absorption tests may also be required.

The costs of testing installed materials may be included—in different ways—as project overhead. One method is separately itemizing the costs for each individual test; or, a fixed allowance can be made based on the size and type of project. Budget costs derived from this second method are shown in Figure 9.18. A third approach is to include a percentage of total project costs.

Temporary Services

Required temporary services may or may not be included in the specifications. A typical statement in the specifications is: "Contractor shall supply all material, labor, equipment, tools, utilities, and other items and services required for the proper and timely performance of the work and completion of the project." As far as the owner and designer are concerned, such a statement eliminates a great deal of ambiguity. To the estimator, this means many extra items that must be estimated. Temporary utilities, such as heat, light, power, and water, are a major consideration. The estimator must not only account for anticipated monthly (or time-related) costs, but should also be sure that installation and removal costs are included, whether by the appropriate subcontractor or the general contractor.

The time of year may also have an impact on the cost of temporary services. Snow removal costs must be anticipated in some climates. If construction begins during a wet season, or there is a high ground water table, then dewatering may be necessary. Typically, the boring logs will give the contractor a feel for probable ground water elevation. Logs should be examined for the time of year in which the borings were taken. If high infiltration rates are expected, wellpoints may be required. The pumping allowance is usually priced from an analysis of the expected duration and volume of water. This information dictates pump size, labor to install, and power to operate the pumps. This can be an expensive item, since the pumps may have to operate 168 hours per week during certain phases of construction.

An office trailer and/or storage trailers or containers are usually required and included in the specifications. Even if these items are owned by the contractor, costs should still be allocated to the job as part of project overhead. Telephone, utility, and temporary toilet facilities are other costs in this category.

Depending on the location and local environment, security measures may be required. In addition to security personnel or guard dogs, fences, gates, special lighting, and alarms may also be needed. A guard shack with heat, power, and telephone can be an expensive temporary cost.

Temporary Construction

Temporary construction may also involve many items that are not specified in the construction documents. Temporary partitions, doors, fences, and barricades may be required to delineate or isolate portions of the building or site. In addition to these items, railings, catwalks, or safety nets may also be necessary for the protection of the workers. Depending on the project size, an OSHA representative may visit the site to ensure that all such safety precautions are being observed.

Ramps, temporary stairs, and ladders are often necessary during construction for access between floors. When the permanent stairs are installed, temporary wood fillers are needed in metal pan treads until the concrete fill is placed. Workers will almost always use a new, permanent elevator for access throughout the building. While this use is often restricted, precautionary measures must be taken to protect the doors and cab. Invariably, some damage occurs. Protection of any and all finished surfaces throughout the course of the project must be priced and included in the estimate.

Job Cleanup

An amount should always be carried in the estimate for cleanup of the grounds and the building, both during the construction process and upon completion. Cleanup can be broken down into three basic categories, which can be estimated separately:

- Continuous (daily or otherwise) cleaning of the building and site
- Rubbish handling and removal
- Final cleanup

Costs for continuous cleaning can be included as an allowance, or estimated by required labor-hours. Rubbish handling should include barrels, a trash chute if necessary, dumpster rental, and disposal fees. These fees vary depending on the project. A permit may also be required. Costs for final cleanup should be based on past projects and may include subcontract costs for items such as cleaning windows and waxing floors. Included in the costs for final cleanup may be an allowance for repair and minor damage to finished work.

Miscellaneous General Conditions

Many other items must be taken into account when costs are being determined for project overhead. Among the major considerations are:

1. *Scaffolding or Swing Staging* – It is important to determine who is responsible for rental, erection, and dismantling of scaffolding. If a subcontractor is responsible, it may be necessary to leave the

scaffolding in place long enough for use by other trades. Scaffolding is priced by the section or per hundred square feet.

2. *Small Tools* – An allowance, based on past experience, should be carried for small tools. It should cover hand tools as well as small power tools for use by workers on the general contractor's payroll. Small tools have a habit of "walking," and a certain amount of replacement is necessary. Special tools like magnetic drills may be required for specific tasks.

3. *Permits* – Various types of permits may be required depending on local codes and regulations. Following are some examples:

    a. General building permit
    b. Subtrade permits (mechanical, electrical, etc.)
    c. Street use permit
    d. Sidewalk use permit
    e. Permit to allow work on Sundays
    f. Rubbish burning permit (if allowed)
    g. Blasting permit

    Both the necessity of the permit and the responsibility for acquiring it must be determined. If the work is being done in an unfamiliar location, local building officials should be consulted regarding unusual or unknown requirements.

4. *Insurance* – Insurance coverage for each project and locality—above and beyond normal, required operating insurance—should be reviewed to ensure that it is adequate. The contract documents will often specify certain required policy limits. The need for specific policies or riders should be anticipated (for example, fire or XCU— explosion collapse, underground).

Other items commonly included in project overhead are: photographs, job signs, sample panels and materials for owner/architect approval, and an allowance for replacement of broken glass. For some materials, such as imported goods or custom-fabricated items, both shipping costs and off-site storage fees can be expected. An allowance should be included for anticipated costs pertaining to punchlist items. These costs are likely to be based on past experience.

Some project overhead costs can be calculated at the beginning of the estimate. Others will be included as the estimating process proceeds. Still other costs are estimated last, since they depend on the total cost and duration of the project. Because many of the overhead items are not directly specified, the estimator must use experience and visualize the construction process to ensure that all requirements are met. It is not important when or where these items are included, but that they *are* included. One contractor may list certain costs as project overhead, while another contractor would allocate the same costs (and responsibility) to a subcontractor. Either way, the costs are recorded in the estimate.

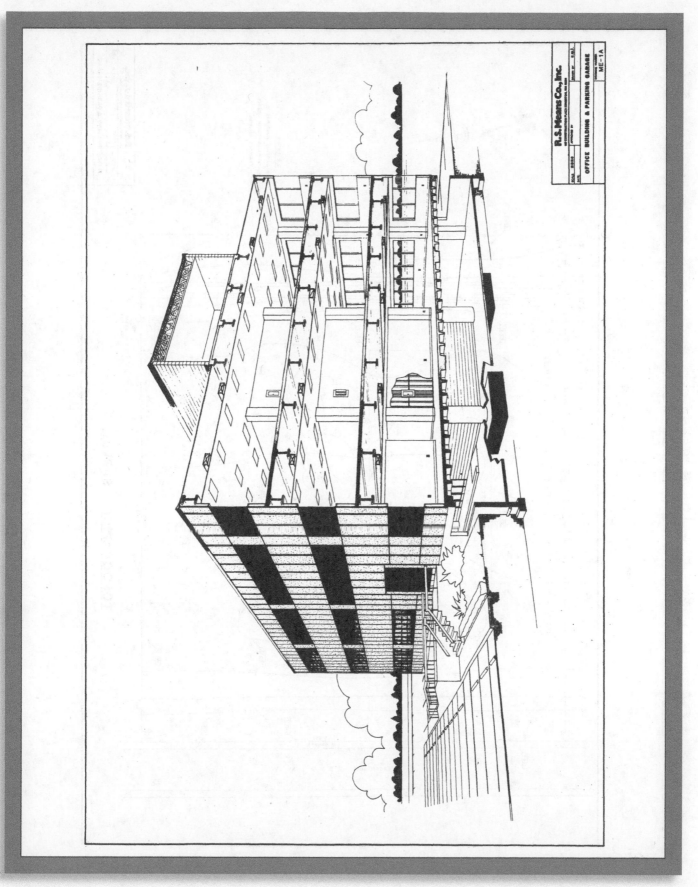

Figure 9.1

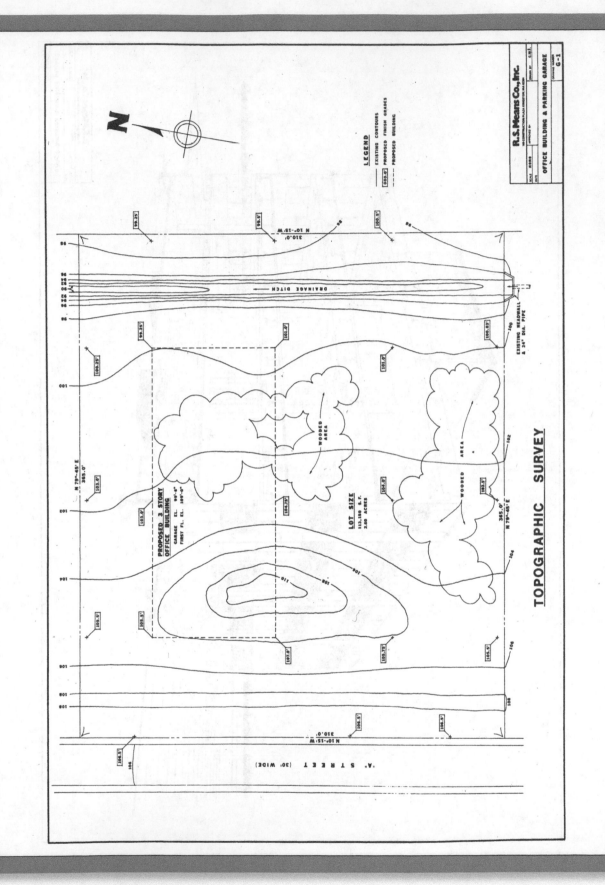

TOPOGRAPHIC SURVEY

Figure 9.2

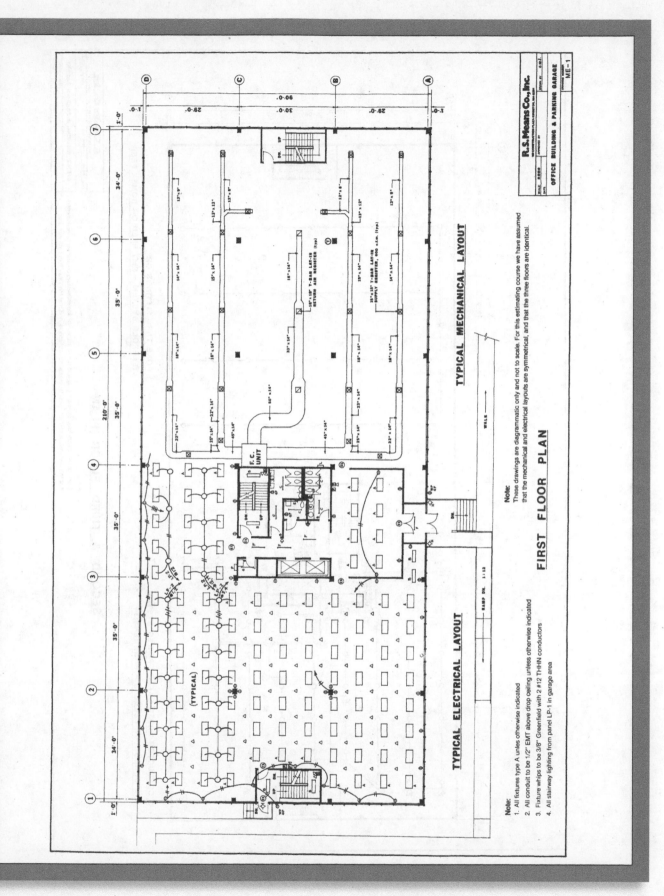

TYPICAL MECHANICAL LAYOUT

FIRST FLOOR PLAN

TYPICAL ELECTRICAL LAYOUT

Note:
These drawings are diagrammatic only and not to scale. For this estimating course we have assumed that the mechanical and electrical layouts are symmetrical, and that the three floors are identical.

Note:
1. All fixtures type A unless otherwise indicated
2. All conduit to be 1/2" EMT above drop ceiling unless otherwise indicated
3. Fixture whips to be 3/8" Greenfield with 2 #12 TH-HN conductors
4. All stairway lighting from panel LP-1 in garage area

OFFICE BUILDING & PARKING GARAGE
ME-1

Figure 9.12

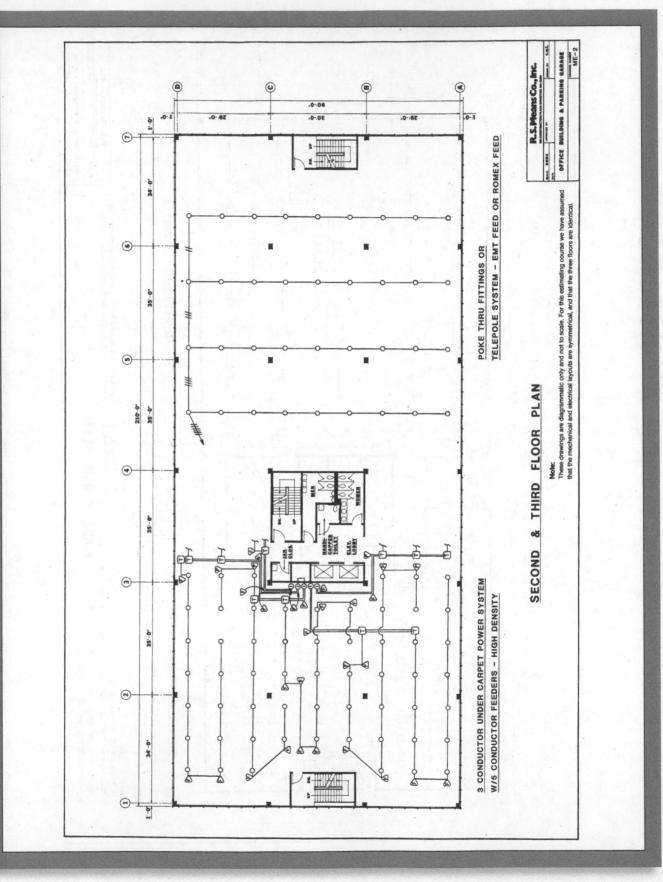

SECOND & THIRD FLOOR PLAN

Figure 9.13

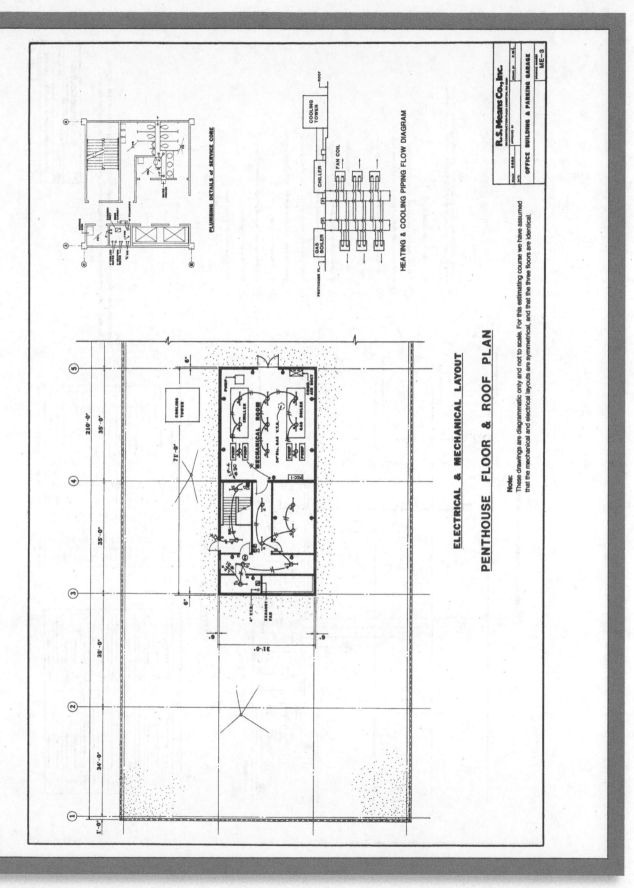

Figure 9.14

141

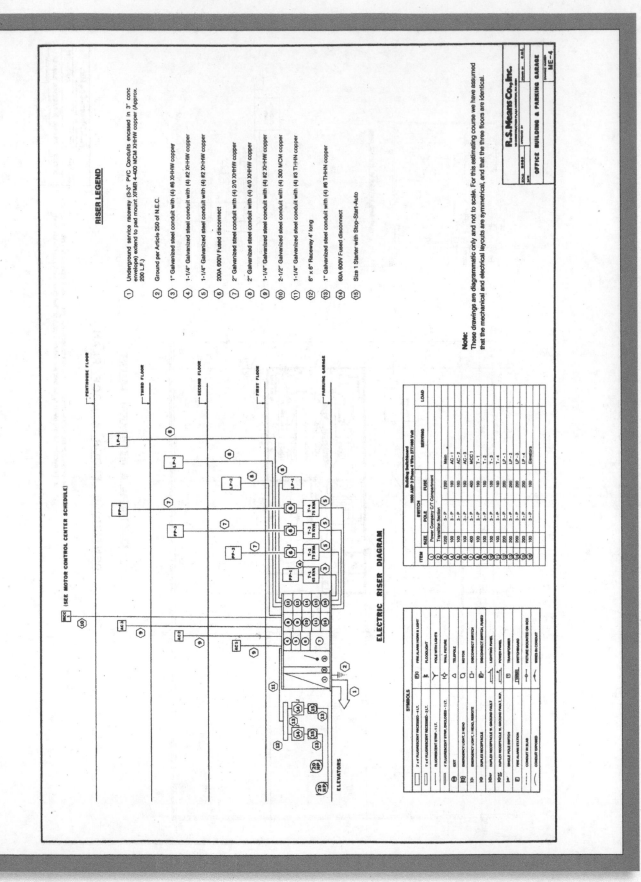

Figure 9.15

PROJECT OVERHEAD SUMMARY

SHEET NO.

PROJECT: Office Building

ESTIMATE NO:

LOCATION: ARCHITECT: DATE:

QUANTITIES BY: PRICES BY: EXTENSIONS BY: CHECKED BY:

DESCRIPTION	QUANTITY	UNIT	MATERIAL/EQUIP.		LABOR		TOTAL COST	
			UNIT	TOTAL	UNIT	TOTAL	UNIT	TOTAL
Job Organization: Superintendent								
Project Manager								
Timekeeper & Material Clerk								
Clerical								
Safety, Watchman & First Aid								
Travel Expense: Superintendent								
Project Manager								
Engineering: Layout								
Inspection / Quantities								
Drawings								
CPM Schedule								
Testing: Soil								
Materials								
Structural								
Equipment: Cranes								
Concrete Pump, Conveyor, Etc.								
Elevators, Hoists								
Freight & Hauling								
Loading, Unloading, Erecting, Etc.								
Maintenance								
Pumping								
Scaffolding								
Small Power Equipment / Tools								
Field Offices: Job Office, Trailer								
Architect / Owner's Office								
Temporary Telephones								
Utilities								
Temporary Toilets								
Storage Areas & Sheds								
Temporary Utilities: Heat								
Light & Power								
PAGE TOTALS								

Figure 9.16

DESCRIPTION	QUANTITY	UNIT	MATERIAL/EQUIP.		LABOR		TOTAL COST	
			UNIT	TOTAL	UNIT	TOTAL	UNIT	TOTAL
Totals Brought Forward								
Winter Protection: Temp. Heat/Protection								
Snow Plowing								
Thawing Materials								
Temporary Roads								
Signs & Barricades: Site Sign								
Temporary Fences								
Temporary Stairs, Ladders & Floors								
Photographs								
Clean Up								
Dumpster								
Final Clean Up								
Continuous - One Laborer								
Punch List								
Permits: Building								
Misc.								
Insurance: Builders Risk - Additional Rider								
Owner's Protective Liability								
Umbrella								
Unemployment Ins. & Social Security								
(See Estimate Summary)								
Bonds								
Performance (See Estimate Summary)								
Material & Equipment								
Main Office Expense (See Estimate Summary)								
Special Items								
Totals:								

Figure 9.17

01 32 Construction Progress Documentation

01 32 13 - Scheduling of work

01 32 13.50 Scheduling	Crew	Daily Output	Labor-Hours	Unit	Material	2007 Bare Costs Labor	Equipment	Total	Total Incl O&P
0010 SCHEDULING									
0020 Critical path, as % of architectural fee, minimum				%					.50%
0100 Maximum				"					1%
0300 Computer-update, micro, no plots, minimum				Ea.				455	500
0400 Including plots, maximum				"				1,455	1,600
0600 Rule of thumb, CPM scheduling, small job ($10 Million)				Job					.05%
0650 Large job ($50 Million +)									.03%
0700 Including cost control, small job									.08%
0750 Large job									.04%

01 32 33 - Photographic Documentation

01 32 33.50 Photographs	Crew	Daily Output	Labor-Hours	Unit	Material	2007 Bare Costs Labor	Equipment	Total	Total Incl O&P
0010 PHOTOGRAPHS									
0020 8" x 10", 4 shots, 2 prints ea., std. mounting				Set	450			450	495
0100 Hinged linen mounts					530			530	580
0200 8" x 10", 4 shots, 2 prints each, in color					455			455	505
0300 For I.D. slugs, add to all above					5.30			5.30	5.85
0500 Aerial photos, initial fly-over, 6 shots, 1 print ea., 8" x 10"					790			790	870
0550 11" x 14" prints					920			920	1,025
0600 16" x 20" prints					1,150			1,150	1,250
0700 For full color prints, add					40%				40%
0750 Add for traffic control area					294			294	325
0900 For over 30 miles from airport, add per				Mile	5.30			5.30	5.85
1000 Vertical photography, 4 to 6 shots with									
1010 different scales, 1 print each				Set	1,100			1,100	1,200
1500 Time lapse equipment, camera and projector, buy					3,775			3,775	4,175
1550 Rent per month					565			565	620
1700 Cameraman and film, including processing, B.&W.				Day	1,375			1,375	1,525
1720 Color				"	1,375			1,375	1,525

01 41 Regulatory Requirements

01 41 26 - Permits

01 41 26.50 Permits	Crew	Daily Output	Labor-Hours	Unit	Material	2007 Bare Costs Labor	Equipment	Total	Total Incl O&P
0010 PERMITS									
0020 Rule of thumb, most cities, minimum				Job					.50%
0100 Maximum				"					2%

01 45 Quality Control

01 45 23 - Testing and Inspecting Services

01 45 23.50 Testing	Crew	Daily Output	Labor-Hours	Unit	Material	2007 Bare Costs Labor	Equipment	Total	Total Incl O&P
0010 TESTING and Inspecting Services									
0015 For concrete building costing $1,000,000, minimum				Project				4,725	5,200
0020 Maximum								38,000	41,800
0050 Steel building, minimum								4,727	5,200
0070 Maximum								14,818	16,300
0100 For building costing, $10,000,000, minimum								30,091	33,100
0150 Maximum								48,182	53,000
0200 Asphalt testing, compressive strength Marshall stability, set of 3				Ea.				145	165
0220 Density, set of 3								86	95
0250 Extraction, individual tests on sample								136	150

Figure 9.18

Credit: *Means Building Construction Cost Data 2007*

Slabs

Concrete slabs can be divided into two categories: slabs on grade (including sidewalks and concrete paving) and elevated slabs. Several items must be considered when estimating slabs on grade:

- Granular base
- Fine grading
- Vapor barrier
- Edge forms and bulkheads
- Expansion joints
- Contraction joints
- Screeds
- Welded wire fabric and reinforcing
- Concrete material
- Finish and topping
- Curing and hardeners
- Control joints
- Concrete "outs"
- Haunches
- Drops

Quantities for most of the items above can be derived from basic dimensions: length, width, and slab thickness. When these quantities are calculated, some basic rules can be used:

- Allow about 25% compaction for granular base.
- Allow 10% overlap for vapor barrier and welded wire fabric.
- Allow 5% extra concrete for waste.
- No deductions should be made for columns or "outs" under 10 SF.
- If screeds are separated from forming and placing costs, figure 1 LF of screed per 10 SF of finish area.

Elevated slabs, while similar to slabs on grade, require special consideration. For all types of elevated slabs, edge forms should be estimated carefully. Stair towers, elevator shafts, and utility chases all require edge forms. The appropriate quantities for each of these "outs" should be deducted from the concrete volume and finish area. Edge form materials and installation will vary depending on the slab type and thickness. Special consideration should be given to pipe chases that require concrete placement after the pipes are in position. This is done in order to maintain the fireproofing integrity. Mixing and placing by hand may be necessary.

When estimating beam-supported slabs, beams should not normally be deducted from the floor form area. The extra costs for bracing and framing at the beams will account for the difference. When concrete for the beams and slab is placed at the same time, the placement costs will be the same for both.

Slab placement costs should be increased based on the beam volume. Separate costs for beam concrete placement should not be included. For hung slabs, separation of costs for beams and slab is even more difficult. These costs may be treated as one item.

Metal decking for elevated slabs is usually installed by the steel erector. When determining concrete quantities, measure the slab thickness to the middle of the corrugation.

Cast-in-place concrete joist and dome (or waffle) slabs are more difficult to estimate. Pans and domes are shored with either open or closed deck forming. For concrete joist systems, quantities for concrete can be calculated from Figure 9.29. Volume of beams must be added separately. Volume of waffle slabs, because of the varying sizes of solid column heads, should be estimated differently. Total volume from top of slab to bottom of pan should be calculated. The volume of the actual number of domes (voids) is then deducted. Volumes for standard domes are listed in Figure 9.30.

Stairs

When taking off cast-in-place stairs, a careful investigation should be made to ensure that all inserts, such as railing pockets, anchors, nosings, and reinforcing steel, have been included. Also indicate on the quantity sheet any special tread finishes; these can sometimes represent a considerable cost.

Stairs cast on fill are most easily estimated and recorded by the square foot. The calculation is slant length multiplied by width. Because of the commonly used ratios adopted by designers, this method works for high or low risers with wide or narrow treads. Shored cast-in-place stairs can be estimated in the same way—slant length times width.

Treads and landings of prefabricated metal pan stairs should be identified on the takeoff sheet. The cost of hand-placing the fill concrete should be included in Division 3.

For all cast-in-place, formed concrete work, the following factors can affect the number of pans to rent, the SF of forms to build, the number of shores to provide, etc.:

- Placement rate of concrete for crew
- Concrete pump or crane and bucket rental cost per pour
- Finishing rate for concrete finishers
- Curing time before stripping and re-shoring
- Forming and stripping time
- Number of reuses of forms

Experience (or consultation with an experienced superintendent) will tell the estimator which combination of these six items will limit the quantity of forms to be built, or the number of pans to be rented.

Precast Concrete

Typical precast units are estimated and purchased by the square foot or linear foot. The manufacturer will most often deliver precast units. Costs for delivery are sometimes separate from the purchase price and must be included in the estimate. Most manufacturers also have the capability to erect these units, and, because of experience, are best suited to the task.

Each type and size of precast unit should be separated on the quantity or estimate sheet. Typical units are shown in Figure 9.31. The estimator should verify the availability of specified units with local suppliers. For types that are seldom used, casting beds may have to be custom built, thus increasing cost.

Connections and joint requirements will vary with the type of unit, from welding or bolting embedded steel to simple grouting. Toppings must also be included, whether lightweight or regular concrete on floors, or insulating concrete for roofs. The specifications and construction details must be examined carefully.

03 30 Cast-In-Place Concrete

03 30 53 – Miscellaneous Cast-In-Place Concrete

03 30 53.40 Concrete In Place		Crew	Daily Output	Labor-Hours	Unit	Material	Labor	2007 Bare Costs Equipment	Total	Total Incl O&P
6350	16' high	C-14D	91	2.198	C.Y.	149	80.50	8	237.50	298
6800	Stairs, not including safety treads, free standing, 3'-6" wide	C-14H	83	.578	LF Nose	5.70	21	.26	26.96	39
6850	Cast on ground	↓	125	.384	"	4.52	13.95	.17	18.64	26.50
7000	Stair landings, free standing		200	.240	S.F.	4.63	8.70	.11	13.44	18.75
7050	Cast on ground	↓	475	.101	"	3.45	3.67	.05	7.17	9.55

03 31 Structural Concrete

03 31 05 – Normal Weight Structural Concrete

03 31 05.30 Concrete, Field Mix

			Crew	Daily Output	Labor-Hours	Unit	Material	Labor	Equipment	Total	Total Incl O&P
0010	**CONCRETE, FIELD MIX**	R033105-65									
0015	FOB forms 2250 psi					C.Y.	92.50			92.50	102
0020	3000 psi					"	96.50			96.50	106

03 31 05.35 Normal Weight Concrete, Ready Mix

			Crew	Daily Output	Labor-Hours	Unit	Material	Labor	Equipment	Total	Total Incl O&P
0010	**NORMAL WEIGHT CONCRETE, READY MIX**	R033105-10									
0012	Includes local aggregate, sand, portland cement, and water										
0015	Excludes all additives and treatments	R033105-20									
0020	2000 psi					C.Y.	99.50			99.50	110
0100	2500 psi	R033105-30					101			101	111
0150	3000 psi	CN					104			104	114
0200	3500 psi	R033105-40					106			106	116
0300	4000 psi						108			108	119
0350	4500 psi	R033105-50					110			110	121
0400	5000 psi	CN					114			114	125
0411	6000 psi						130			130	143
0412	8000 psi						212			212	233
0413	10,000 psi						300			300	330
0414	12,000 psi						365			365	400
1000	For high early strength cement, add						10%				
1010	For structural lightweight with regular sand, add						25%				
2000	For all lightweight aggregate, add					↓	45%				

03 31 05.70 Placing Concrete

			Crew	Daily Output	Labor-Hours	Unit	Material	Labor	Equipment	Total	Total Incl O&P
0010	**PLACING CONCRETE**	R033105-70									
0020	Includes labor and equipment to place and vibrate										
0050	Beams, elevated, small beams, pumped		C-20	60	1.067	C.Y.		33	12.50	45.50	65
0100	With crane and bucket		C-7	45	1.600			50	25	75	105
0200	Large beams, pumped		C-20	90	.711			22	8.35	30.35	43
0250	With crane and bucket		C-7	65	1.108			35	17.25	52.25	72.50
0400	Columns, square or round, 12" thick, pumped		C-20	60	1.067			33	12.50	45.50	65
0450	With crane and bucket		C-7	40	1.800			56.50	28	84.50	118
0600	18" thick, pumped		C-20	90	.711			22	8.35	30.35	43
0650	With crane and bucket		C-7	55	1.309			41	20.50	61.50	85.50
0800	24" thick, pumped		C-20	92	.696			21.50	8.15	29.65	42
0850	With crane and bucket		C-7	70	1.029			32.50	16	48.50	67
1000	36" thick, pumped		C-20	140	.457			14.20	5.35	19.55	28
1050	With crane and bucket		C-7	100	.720			22.50	11.20	33.70	47
1400	Elevated slabs, less than 6" thick, pumped		C-20	140	.457			14.20	5.35	19.55	28
1450	With crane and bucket		C-7	95	.758			24	11.80	35.80	49.50
1500	6" to 10" thick, pumped		C-20	160	.400			12.40	4.70	17.10	24.50
1550	With crane and bucket		C-7	110	.655			20.50	10.20	30.70	42.50
1600	Slabs over 10" thick, pumped		C-20	180	.356	↓		11.05	4.17	15.22	21.50

Figure 9.21

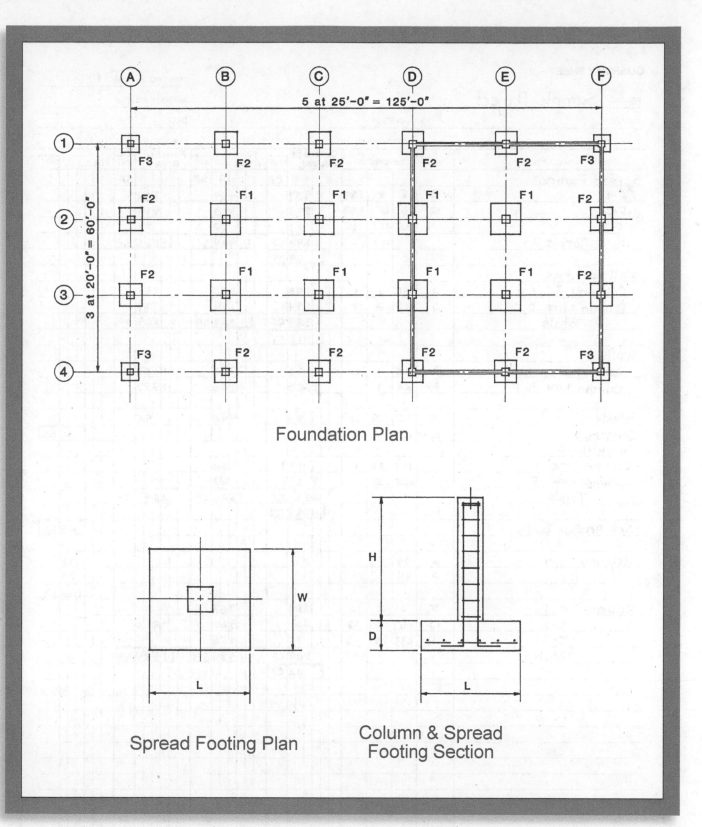

Foundation Plan

Spread Footing Plan

Column & Spread
Footing Section

Figure 9.22

QUANTITY SHEET

PROJECT **Sample Project**

ESTIMATE NO.

LOCATION	ARCHITECT	DATE
TAKE OFF BY	EXTENSIONS BY:	CHECKED BY:

DESCRIPTION	NO.	DIMENSIONS			Concrete Volume	UNIT	Form Area	UNIT	Finished Area	UNIT	Misc.	UNIT
Spread Footings						CF		SF		SF		
F-1	8	8	8	1.83	937		469		512			
F-2	12	6	6	1.33	575		383		432			
F-3	4	4.5	4.5	1	81		72		81			
Totals					1593	CF	924	SF	1025	SF		
					59	CY						
Wall Footings												
Column Line 1,4	2	38.75	2	1	155		155		155			
Column Line D,F	2	43.5	2	1	174		174		174			
Totals					329	CF	329	SF	329	SF		
					13	CY						
Walls												
Column Line 1,4	2	51.5	1	11	1133		2266		1133			
Column Line D,F	2	61.5	1	11	1353		2706		1353			
Pilasters	10	1.5	1.5	11	83		440		55			
Corners	10 × 8										80	Ea.
Brick Shelf												
Column Line 1		51.5	.33	1	(17)		52					
Column Line F		61.5	.33	1	(21)		62					
Totals					2531	CF	5526	SF	2541	SF		
					94	CY						
Set Anchor Bolts											48	Ea.
Keyway	2	39									78	
	2	44									88	
											166	LF
Columns C-1	8	2	2	12	384		768					
C-2	12	1.67	1.67	12	400		960		ALL			
C-3	4	1.33	1.33	12	85		256		↓			
Totals					869	CF	1984	SF	1984	SF		
					32	CY						

Figure 9.23

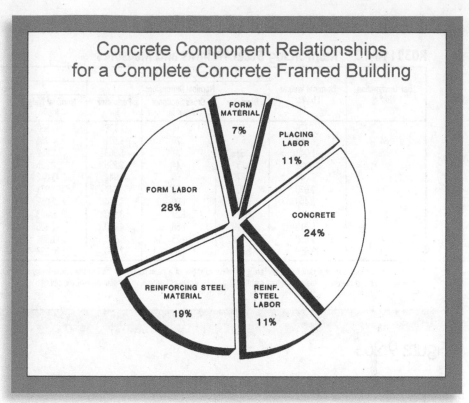

Concrete Component Relationships for a Complete Concrete Framed Building

- FORM MATERIAL 7%
- PLACING LABOR 11%
- CONCRETE 24%
- FORM LABOR 28%
- REINFORCING STEEL MATERIAL 19%
- REINF. STEEL LABOR 11%

Figure 9.24

R031113-40   Forms for Reinforced Concrete

Design Economy

Avoid many sizes in proportioning beams and columns.

From story to story avoid changing column dimensions. Gain strength by adding steel or using a richer mix. If a change in size of column is necessary, vary one dimension only to minimize form alterations. Keep beams and columns the same width.

From floor to floor in a multi-story building vary beam depth, not width, as that will leave slab panel form unchanged. It is cheaper to vary the strength of a beam from floor to floor by means of steel area than by 2" changes in either width or depth.

Cost Factors

Material includes the cost of lumber, cost of rent for metal pans or forms if used, nails, form ties, form oil, bolts and accessories.

Labor includes the cost of carpenters to make up, erect, remove and repair, plus common labor to clean and move. Having carpenters remove forms minimizes repairs.

Improper alignment and condition of forms will increase finishing cost. When forms are heavily oiled, concrete surfaces must be neutralized before finishing. Special curing compounds will cause spillages to spall off in first frost. Gang forming methods will reduce costs on large projects.

Materials Used

Boards are seldom used unless their architectural finish is required. Generally, steel, fiberglass and plywood are used for contact surfaces. Labor on plywood is 10% less than with boards. The plywood is backed up with 2 x 4's at 12" to 32" O.C. Walers are generally 2 - 2 x 4's. Column forms are held together with steel yokes or bands. Shoring is with adjustable shoring or scaffolding for high ceilings.

Reuse

Floor and column forms can be reused four or possibly five times without excessive repair. Remember to allow for 10% waste on each reuse.

When modular sized wall forms are made, up to twenty uses can be expected with exterior plyform.

When forms are reused, the cost to erect, strip, clean and move will not be affected. 10% replacement of lumber should be included and about one hour of carpenter time for repairs on each reuse per 100 S.F.

The reuse cost for certain accessory items normally rented on a monthly basis will be lower than the cost for the first use.

After fifth use, new material required plus time needed for repair prevent form cost from dropping further and it may go up. Much depends on care in stripping, the number of special bays, changes in beam or column sizes and other factors.

Costs for multiple use of formwork may be developed as follows:

2 Uses	3 Uses	4 Uses
$\frac{(\text{1st Use} + \text{Reuse})}{2}$ = avg. cost/2 uses	$\frac{(\text{1st Use} + \text{2 Reuse})}{3}$ = avg. cost/3 uses	$\frac{(\text{1st use} + \text{3 Reuse})}{4}$ = avg. cost/4 uses

Figure 9.25

Credit: Means Building Construction Cost Data 2007

R032110-10   Reinforcing Steel Weights and Measures

Bar Designation No.**	Nominal Weight Lb./Ft.	U.S. Customary Units			SI Units			
		Nominal Dimensions*			Nominal Dimensions*			
		Diameter in.	Cross Sectional Area, in.^{2}	Perimeter in.	Nominal Weight kg/m	Diameter mm	Cross Sectional Area, cm^2	Perimeter mm
3	.376	.375	.11	1.178	.560	9.52	.71	29.9
4	.668	.500	.20	1.571	.994	12.70	1.29	39.9
5	1.043	.625	.31	1.963	1.552	15.88	2.00	49.9
6	1.502	.750	.44	2.356	2.235	19.05	2.84	59.8
7	2.044	.875	.60	2.749	3.042	22.22	3.87	69.8
8	2.670	1.000	.79	3.142	3.973	25.40	5.10	79.8
9	3.400	1.128	1.00	3.544	5.059	28.65	6.45	90.0
10	4.303	1.270	1.27	3.990	6.403	32.26	8.19	101.4
11	5.313	1.410	1.56	4.430	7.906	35.81	10.06	112.5
14	7.650	1.693	2.25	5.320	11.384	43.00	14.52	135.1
18	13.600	2.257	4.00	7.090	20.238	57.33	25.81	180.1

* The nominal dimensions of a deformed bar are equivalent to those of a plain round bar having the same weight per foot as the deformed bar.
** Bar numbers are based on the number of eighths of an inch included in the nominal diameter of the bars.

Figure 9.26a

Credit: *Means Building Construction Cost Data 2007*

R032205-30   Common Stock Styles of Welded Wire Fabric

This table provides some of the basic specifications, sizes, and weights of welded wire fabric used for reinforcing concrete.

	New Designation	Old Designation		Steel Area per Foot				Approximate Weight per 100 S.F.	
	Spacing — Cross Sectional Area (in.) — (Sq. in. 100)	Spacing — Wire Gauge (in.) — (AS & W)		Longitudinal		Transverse			
				in.	cm	in.	cm	lbs	kg
Rolls	6 x 6 — W1.4 x W1.4	6 x 6 — 10 x 10		.028	.071	.028	.071	21	9.53
	6 x 6 — W2.0 x W2.0	6 x 6 — 8 x 8	1	.040	.102	.040	.102	29	13.15
	6 x 6 — W2.9 x W2.9	6 x 6 — 6 x 6		.058	.147	.058	.147	42	19.05
	6 x 6 — W4.0 x W4.0	6 x 6 — 4 x 4		.080	.203	.080	.203	58	26.91
	4 x 4 — W1.4 x W1.4	4 x 4 — 10 x 10		.042	.107	.042	.107	31	14.06
	4 x 4 — W2.0 x W2.0	4 x 4 — 8 x 8	1	.060	.152	.060	.152	43	19.50
	4 x 4 — W2.9 x W2.9	4 x 4 — 6 x 6		.087	.227	.087	.227	62	28.12
	4 x 4 — W4.0 x W4.0	4 x 4 — 4 x 4		.120	.305	.120	.305	85	38.56
Sheets	6 x 6 — W2.9 x W2.9	6 x 6 — 6 x 6		.058	.147	.058	.147	42	19.05
	6 x 6 — W4.0 x W4.0	6 x 6 — 4 x 4		.080	.203	.080	.203	58	26.31
	6 x 6 — W5.5 x W5.5	6 x 6 — 2 x 2	2	.110	.279	.110	.279	80	36.29
	4 x 4 — W1.4 x W1.4	4 x 4 — 4 x 4		.120	.305	.120	.305	85	38.56

NOTES: 1. Exact W—number size for 8 gauge is W2.1
2. Exact W—number size for 2 gauge is W5.4

Figure 9.26b

Credit: *Means Building Construction Cost Data 2007*

Footing Schedule			
Ident.	No.	Size	Reinforcing
F-1	8	8'-0" x 8'0" x 1'-10"	9-#6 e.w.
F-2	12	6'-0" x 6'0" x 1'-4"	8-#5 e.w.
F-3	4	4'-6" x 4'6" x 1'-0"	5-#5 e.w.

Column Schedule			
Ident.	No.	Size	Reinforcing
C-1	8	24" x 24" x 12'	8-#11 ties #4 @ 22"
C-2	12	20" x 20" x 12'	8-#9 ties #3 @ 18"
C-3	4	16" x 16" x 12'	8-#9 ties #3 @ 16"

Beam Schedule			
Ident.	No.	Size	Reinforcing
B-1 Ext.	1	1' x 2'-6" x 60'	See Re-Bar Schedule
B-2 Ext.	2	1' x 2'-4" x 75'	
B-3 Int.	2	8" x 1'-6" x 60'	
B-4 Int.	2	6" x 1'-6" x 125'	

Figure 9.27

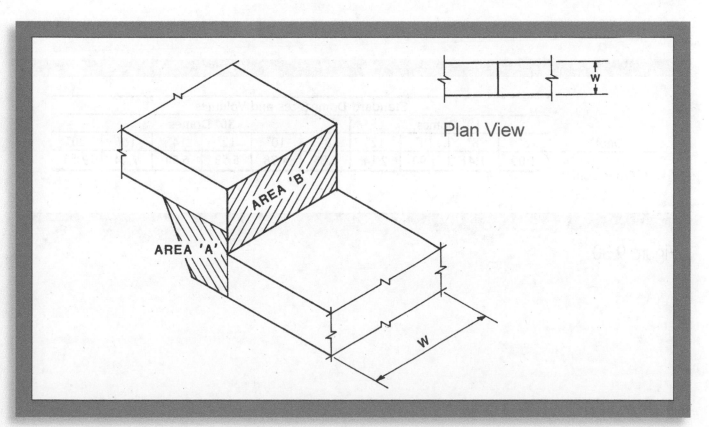

Plan View

AREA 'B'

AREA 'A'

W

Figure 9.28

Concrete Quantities (CF Concrete/SF Floor) for Single & Multiple Span Concrete Joist Construction						
Width						
	20" Forms			30" Forms		
Depth (Rib/Slab)	5" Rib 25" O.C.	6" Rib 26" O.C.	7" Rib 27" O.C.	5" Rib 35" O.C.	6" Rib 36" O.C.	7" Rib 37" O.C.
8"/3"	0.40	0.42	—	0.36	0.37	—
8"/4½"	0.53	0.55	—	0.48	0.50	—
10"/3"	0.45	0.47	—	0.39	0.41	—
10"/4½"	0.57	0.61	—	0.51	0.53	—
12"/3"	0.49	0.52	—	0.42	0.45	—
12"/4½"	0.62	0.65	—	0.55	0.57	—
14"/3"	0.54	0.57	—	0.45	0.48	—
14"/4½"	0.66	0.69	—	0.58	0.61	—
16"/3"	—	0.63	0.66	—	0.52	0.55
16"/4½"	—	0.75	0.79	—	0.65	0.68
20"/3"	—	0.75	0.79	—	0.61	0.64
20"/4½"	—	0.87	0.91	—	0.74	0.77

Figure 9.29

Standard Dome Sizes and Volumes									
	19" Domes				30" Domes					
Depth	6"	8"	10"	12"	8"	10"	12"	14"	16"	20"
Volume (CF per Dome)	1.09	1.41	1.90	2.14	3.85	4.78	5.53	6.54	7.44	9.16

Figure 9.30

R034105-30   Prestressed Precast Concrete Structural Units

Type	Location	Depth	Span in Ft.		Live Load Lb. per S.F.
Double Tee	Floor	28" to 34"	60 to 80		50 to 80
	Roof	12" to 24"	30 to 50		40
	Wall	Width 8'	Up to 55' high		Wind
Multiple Tee	Roof	8" to 12"	15 to 40		40
	Floor	8" to 12"	15 to 30		100
Plank or	Roof or Floor		Roof	Floor	40 for Roof
		4"	13	12	
		6"	22	18	
		8"	26	25	
		10"	33	29	100 for Floor
		12"	42	32	
Single Tee	Roof	28"	40		
		32"	80		
		36"	100		40
		48"	120		
AASHO Girder	Bridges	Type 4	100		
		5	110		Highway
		6	125		
Box Beam	Bridges	15"	40		
		27"	to		Highway
		33"	100		

The majority of precast projects today utilize double tees rather than single tees because of speed and ease of installation. As a result casting beds at manufacturing plants are normally formed for double tees. Single tee projects will therefore require an initial set up charge to be spread over the individual single tee costs.

For floors, a 2" to 3" topping is field cast over the shapes. For roofs, insulating concrete or rigid insulation is placed over the shapes.

Member lengths up to 40' are standard haul, 40' to 60' require special permits and lengths over 60' must be escorted. Over width and/or over length can add up to 100% on hauling costs.

Large heavy members may require two cranes for lifting which would increase erection costs by about 45%. An eight man crew can install 12 to 20 double tees, or 45 to 70 quad tees or planks per day.

Grouting of connections must also be included.

Several system buildings utilizing precast members are available. Heights can go up to 22 stories for apartment buildings. Optimum design ratio is 3 S.F. of surface to 1 S.F. of floor area.

Figure 9.31

Credit: *Means Building Construction Cost Data 2007*

For the estimate for the sample building project, a number of principles mentioned in the previous sections apply. The most important is common sense, together with the ability to visualize the construction process. Most information for the quantity takeoff would normally be supplied on drawings, details, and in the specifications. Even when all data is provided, however, it is important to be familiar with standard materials and practices.

Concrete work is composed of many components, most of which are based on similar dimensions. It makes sense to organize all takeoff data on a quantity sheet as shown in Figure 9.23. A partial takeoff for Division 3 of the sample project is shown in Figure 9.32. Note that every quantity that is to be priced has been converted to the appropriate units and delineated on the sheet. As the quantities are transferred to the pricing sheets, colored pencil check marks will ensure inclusion of all items. The estimate sheets for Division 3 are shown in Figures 9.33 to 9.37.

Most quantities are derived from the plan in Figure 9.5a. Heights are determined from the building section in Figure 9.10. Dimensioned details are provided in complete plans and specifications. The spread footing quantities are easily obtained from the drawings. Care must be exercised when estimating the continuous footings. Note on Figure 9.32 that the complete perimeter is used for a gross quantity and that a deduction is included where the spread footings interrupt the continuous footing. The keyway and dowel supports, as shown in Section A-A in Figure 9.5a, continue across the spread footings. Thus, the linear dimension of the keyway and dowel supports will be greater than that of the continuous footing. (Note that the keyway and dowel supports are included in more than the perimeter footings.) The steps in the perimeter footing between column lines 5 and 6 (Figure 9.5a) are not listed separately due to their insignificant effect on the total cost. This is a judgment call of the type that should only be made based on experience.

Quantities for the perimeter wall are calculated with no deduction for the pilasters. Such a deduction should be more than compensated by extra costs for framing and bracing at the connections. Pilaster formwork should be recorded separately. Costs used for placing the pilaster concrete are the same as for the perimeter wall, because placement for both will occur simultaneously.

Reinforcing quantities are taken from Reference Table R033105-10, Proportionate Quantities for Structural Concrete, in *Means Building Construction Cost Data*. These figures are based on pounds per cubic yard of concrete. When detailed information is available, actual linear quantities and counts of splices and accessories of each size and type should be taken off and priced for greater accuracy. Note that the costs used for welded wire fabric as presented in *Means Building Construction Cost Data* (Figure 9.38) include a 10% overlap so that quantities in this case are derived from actual dimensions.

Concrete material costs for the waffle slab include a cost for high early strength. The 2nd and 3rd floor elevated slabs include percentages for high early strength and for lightweight aggregate. The appropriate percentages are obtained from Figure 9.39 and added to the unit costs before entry on the estimate sheet. Costs for concrete are readily available from local suppliers and should always be verified.

Figures 9.39 (bottom) and 9.40 show costs for complete concrete systems in place. These systems are most often priced per cubic yard of concrete. The costs are based on averages for each type and size of system as listed. These figures provide an excellent checklist to quickly compare costs of similar systems estimated by individual components for possible omissions or duplications. For the sample estimate such a comparison can be made. The component costs for spread footings from the estimate sheets of Division 3 are added:

Spread Footings:

	Bare Costs			
	Material	Labor	Equipment	Total
Formwork	$ 1,185	$5,381		$ 6,566
Reinf.	3,783	2,804		6,587
Concrete	18,720			18,720
Placing		2,178	$65	2,243
Total	$23,688	$10,363	$65	$34,116
$/CY (171 CY)	138.53	60.60	0.38	199.51

Reinforcing costs, separated for the spread footings only, are obtained from the appropriate quantities from Figure 9.32, and the unit costs from Figure 9.35. Comparison to line 3850 in Figure 9.40 verifies that costs for the building's spread footings are good. Similar cross-checks can be made throughout the estimate to help prevent gross errors. Quantities and prices for all components of each of the concrete systems have been separately itemized. By organizing the estimate in this way, costs and quantities can be compared to historical figures and used as a comparison for future projects.

Sample Estimate: Division 3

QUANTITY SHEET

PROJECT: Office Building
LOCATION:
TAKE OFF BY:

Division 3
ARCHITECT:
EXTENSIONS BY:

DESCRIPTION	NO.	DIMENSIONS			Forms	UNIT	Volume	UNIT	Reinf.	UNIT	Finish	UNIT
Spread Footings												
	6	11.5'	11.5'	2'	552	SFCA	1587	CF				
	4	9.5'	9.5'	2'	304		722					
	18	8'	8'	2'	1152		2304					
					2008	SFCA	4613	CF				
Dowel Supports	28 Ea.						171	CY	8892	Lbs.		
									4.45	T		
Continuous Footings												
Perimeter	1	600'	2'	1'	1200	SFCA	1200	CF				
Deduct Spr. Footings	(	8'	2'	1'	-288		-288	)				
Ramp		166'	1.5'	0.67'	498		166					
Stoop		25'	1.5'	0.67'	75		25					
Stairs	2	40'	1.5'	0.67'	240		81					
Core		160'	1.5'	0.67'	480		161					
					2205	SFCA	1345	CF				
Dowel Supports	1031	LF					50	CY	2600	Lbs.		
Keyway	791	LF							1.30	T		
Deduct Stairs / Core												
Walls - Perimeter		600'	1'	10'	12000	SFCA	4020	CF			6000	SF
Pit		42'	0.83'	4'	336		140					
Ramp & Stoop		191'	1'	~4'	1528		512					
Deduct - Doors	(2	10'	1'	10'	-400		-134	)				
Openings	(6	31'	1'	~4.5'	-1674		-561	)				
					11790	SFCA	3977	CF				
Box Openings	466	LF					148	CY	18988	Lbs.		
									9.49	T		
Columns 28" diam.	9	30" diam.	9'		81	LF	398	CF			636	
	1	30" diam.	12'		12		59				94	
					93	LF	457	CY			730	SF
Pilasters 3 Sides	14	2'	9'		673	SFCA	421	CF	8908	Lbs.		
2 Sides	4	1.67'	1.67'	9'	120		100		4.45	T		
					793	SFCA	521	CF	9956	Lbs.	793	SF
							19	CY	4.98	T		
Chamfer	32			9'	288	LF						

Figure 9.32

To download this and other forms in this book, visit www.rsmeans.com/supplement/67303B.asp

Sample Estimate: Division 3

CONSOLIDATED ESTIMATE

PROJECT: Office Building
LOCATION:
TAKE OFF BY: ABC
CLASSIFICATION: Division 3
ARCHITECT:
QUANTITIES BY: ABC
PRICES BY:
EXTENSIONS BY:

ESTIMATE NO:
DATE: Jan-07
CHECKED BY: GHI

PRICES BY: DEF   EXTENSIONS BY: DEF

DESCRIPTION	SOURCE			QUANT	UNIT	MATERIAL		LABOR		EQUIPMENT		SUBCONTRACT		TOTAL	
						COST	TOTAL	COST	TOTAL	COST	TOTAL	COST	TOTAL	COST	TOTAL
Division 3: Concrete															
Form Work															
Spread Footing, job-built lumber	03 11	13.45	5150	2008	SFCA	0.59	1185	2.68	5381						
Dowel Supports	03 11	13.45	6100	18	Ea.	22.00	396	55.50	999						
	03 11	13.45	6150	10	Ea.	26.50	265	65.50	655						
Continuous Footings	03 11	13.45	0150	2205	SFCA	0.86	1896	2.29	5049						
Dowel Supports	03 11	13.45	0500	1031	L.F.	0.84	866	2.22	2289						
Keyway (excl. stairs & core)	03 11	13.45	1500	791	L.F.	0.20	158	0.55	435						
Walls:															
Pit (10")	03 11	13.85	2000	336	SFCA	2.49	837	5.70	1915						
Ramp & Stoops (8")	03 11	13.85	2000	1528	SFCA	2.49	3805	5.70	8710						
Perimeter (12" x 10')	03 11	13.85	2550	9926	SFCA	0.66	6551	4.34	43079						
Deduct Openings															
Garage															
Above Grade															
Box Openings	03 11	13.85	0150	466	L.F.	1.93	899	6.10	2843						
Columns - 28" Diameter, 9@ 9' 1 @12'	03 11	13.25	1850	93	L.F.	11.45	1065	8.90	828						
Pilasters	03 11	13.85	8600	793	SFCA	2.90	2300	6.35	5036						
Subtotals						$	20,223	$	77,218						

Figure 9.33

To download this and other forms in this book, visit www.rsmeans.com/supplement/67303B.asp

Sample Estimate: Division 3

CONSOLIDATED ESTIMATE

PROJECT: Office Building	CLASSIFICATION: Division 3	SHEET NO. 2 of 5
LOCATION:	ARCHITECT:	ESTIMATE NO:
TAKE OFF BY: ABC	QUANTITIES BY: ABC	DATE: Jan-07
	PRICES BY: ABC DEF	CHECKED BY: GHI
	EXTENSIONS BY: DEF	

DESCRIPTION	SOURCE		QUANT	UNIT	MATERIAL COST	MATERIAL TOTAL	LABOR COST	LABOR TOTAL	EQUIPMENT COST	EQUIPMENT TOTAL	SUBCONTRACT COST	SUBCONTRACT TOTAL	TOTAL COST	TOTAL
Division 3: (Cont'd)														
Form Work (Cont'd)														
Chamfer 3/4" wide	03 15 05.12	2200	288	L.F.	0.36	104	0.56	161						
4" Slab Edge Form	03 11 13.65	3000	65	L.F.	0.29	19	1.85	120						
@ Pit & Openings														
Waffle Slab (1st Floor)	03 11 13.35	4500	18220	L.F.	6.45	117519	4.23	77071						
			18900											
Deduct, Stairs			-600											
Shaft			-80											
Opening Edge Forms	03 11 13.35	5000	273	SFCA	3.32	906	9.00	2457						
Perimeter Edge Forms	03 11 13.35	7000	600	SFCA	0.19	114	2.22	1332						
Perimeter Work Deck	03 11 13.35	8000	200	L.F.	13.55	2710	12.35	2470						
Bulkhead Forms	03 11 13.35	6000	5000	L.F.	1.54	7700	3.43	17150						
Reinforcing Steel														
Footings	03 21 10.60	0500	5.75	TON	850.00	4888	630.00	3623						
Walls	03 21 10.60	0700	9.49	TON	850.00	8067	440.00	4176						
Columns	03 21 10.60	0250	9.43	TON	895.00	8440	575.00	5422						
Waffle Slab	03 21 10.60	0400	23.59	TON	950.00	22411	455.00	10733						
					$	172,876	$	124,715						

Figure 9.34

Sample Estimate: Division 3

CONSOLIDATED ESTIMATE

PROJECT: Office Building	CLASSIFICATION: Division 3	ESTIMATE NO:
LOCATION:	ARCHITECT:	DATE: Jan-07
TAKE OFF BY: ABC	QUANTITIES BY: ABC PRICES BY: As Shown EXTENSIONS BY:	CHECKED BY: GHI

DESCRIPTION	SOURCE	QUANT	UNIT	MATERIAL COST DEF	MATERIAL TOTAL	LABOR COST DEF	LABOR TOTAL	EQUIPMENT COST DEF	EQUIPMENT TOTAL	SUBCONTRACT COST	SUBCONTRACT TOTAL	TOTAL COST	TOTAL TOTAL
Division 3: (Cont'd)													
Reinforcing Steel													
WWF: 6 x 6 10/10													
Elevated Slab		37800	S.F.										
Garage Slab		18900											
Penthouse Floor		2100											
Deduct - Stair		-1200											
Elevator		-360											
		57240	S.F.										
Total Welded Wire Fabric (WWF)	03 22 05.50 0100	572.4	C.S.F.	12.75	7298	18.90	10818						
Cast in Place Concrete													
Spreading Footings													
Concrete - Incl. 5% Waste	03 31 05.35 0150	180	C.Y.	104.00	18720								
Placing	03 31 05.70 2600	180	C.Y.			12.10	2178	0.36	65				
Continuous Footings													
Concrete - Incl. 5% Waste	03 31 05.35 0150	52	C.Y.	104.00	5408								
Placing	03 31 05.70 1900	52	C.Y.			12.10	629	0.36	19				
Walls													
Concrete - Incl. 5% Waste	03 31 05.35 0300	233	C.Y.	108.00	25164								
Placing	03 31 05.70 5100	233	C.Y.			18.05	4206	6.85	1596				
Finishing	03 35 29.60 0020	6000	S.F.	0.03	180	0.53	3180						
Columns													
Concrete - Incl. 5% Waste	03 31 05.35 0300	18	C.Y.	108.00	1944								
Placing	03 31 05.70 1000	18	C.Y.			14.20	256	5.35	96				
Finishing	03 35 29.60 0020	730	S.F.	0.03	22	0.53	387						
					$ 58,736		$ 21,654		$ 1,776				

Figure 9.35

To download this and other forms in this book, visit **www.rsmeans.com/supplement/67303B.asp**

04 05 Common Work Results for Masonry

04 05 23 – Masonry Accessories

04 05 23.95 Wall Plugs	Crew	Daily Output	Labor-Hours	Unit	Material	2007 Bare Costs Labor	Equipment	Total	Total Incl O&P
0010 **WALL PLUGS**									
0020 26 ga., galvanized, plain	1 Bric	10.50	.762	C	29.50	29		58.50	76.50
0050 Wood filled	"	10.50	.762	"	100	29		129	154

04 21 Clay Unit Masonry

04 21 13 – Brick Masonry

04 21 13.13 Brick Veneer Masonry

		Crew	Daily Output	Labor-Hours	Unit	Material	2007 Bare Costs Labor	Equipment	Total	Total Incl O&P
0010	**BRICK VENEER MASONRY** R042110-20									
0015	Material costs incl. 3% brick and 25% mortar waste									
0020	Standard, select common, 4" x 2-2/3" x 8" (6.75/S.F.)	D-8	1.50	26.667	M	500	915		1,415	1,950
0050	Red, 4" x 2-2/3" x 8", running bond		1.50	26.667		570	915		1,485	2,025
0100	Full header every 6th course (7.88/S.F.) R042110-50		1.45	27.586		570	945		1,515	2,075
0150	English, full header every 2nd course (10.13/S.F.)		1.40	28.571		570	980		1,550	2,125
0200	Flemish, alternate header every course (9.00/S.F.)		1.40	28.571		570	980		1,550	2,125
0250	Flemish, alt. header every 6th course (7.13/S.F.)		1.45	27.586		570	945		1,515	2,075
0300	Full headers throughout (13.50/S.F.)		1.40	28.571		565	980		1,545	2,125
0350	Rowlock course (13.50/S.F.)		1.35	29.630		565	1,025		1,590	2,175
0400	Rowlock stretcher (4.50/S.F.)		1.40	28.571		575	980		1,555	2,125
0450	Soldier course (6.75/S.F.)		1.40	28.571		570	980		1,550	2,125
0500	Sailor course (4.50/S.F.)		1.30	30.769		575	1,050		1,625	2,225
0601	Buff or gray face, running bond, (6.75/S.F.)		1.50	26.667		570	915		1,485	2,025
0700	Glazed face, 4" x 2-2/3" x 8", running bond		1.40	28.571		1,525	980		2,505	3,175
0750	Full header every 6th course (7.88/S.F.)		1.35	29.630		1,425	1,025		2,450	3,125
1000	Jumbo, 6" x 4" x 12", (3.00/S.F.)		1.30	30.769		1,575	1,050		2,625	3,325
1051	Norman, 4" x 2-2/3" x 12" (4.50/S.F.)		1.45	27.586		980	945		1,925	2,525
1100	Norwegian, 4" x 3-1/5" x 12" (3.75/S.F.)		1.40	28.571		945	980		1,925	2,550
1150	Economy, 4" x 4" x 8" (4.50 per S.F.)		1.40	28.571		920	980		1,900	2,500
1201	Engineer, 4" x 3-1/5" x 8", (5.63/S.F.)		1.45	27.586		580	945		1,525	2,100
1251	Roman, 4" x 2" x 12", (6.00/S.F.)		1.50	26.667		925	915		1,840	2,425
1300	S.C.R. 6" x 2-2/3" x 12" (4.50/S.F.)		1.40	28.571		1,100	980		2,080	2,725
1350	Utility, 4" x 4" x 12" (3.00/S.F.)	↓	1.35	29.630		1,225	1,025		2,250	2,900
1360	For less than truck load lots, add				↓	12			12	13.20
1400	For battered walls, add						30%			
1450	For corbels, add						75%			
1500	For curved walls, add						30%			
1550	For pits and trenches, deduct						20%			
1999	Alternate method of figuring by square foot									
2000	Standard, sel. common, 4" x 2-2/3" x 8", (6.75/S.F.)	D-8	230	.174	S.F.	3.85	5.95		9.80	13.30
2020	Standard, red, 4" x 2-2/3" x 8", running bond (6.75/SF)		220	.182		3.85	6.25		10.10	13.75
2050	Full header every 6th course (7.88/S.F.)		185	.216		4.49	7.40		11.89	16.25
2100	English, full header every 2nd course (10.13/S.F.)		140	.286		5.75	9.80		15.55	21.50
2150	Flemish, alternate header every course (9.00/S.F.)		150	.267		5.10	9.15		14.25	19.55
2200	Flemish, alt. header every 6th course (7.13/S.F.)		205	.195		4.07	6.70		10.77	14.65
2250	Full headers throughout (13.50/S.F.)		105	.381		7.65	13.05		20.70	28.50
2300	Rowlock course (13.50/S.F.)		100	.400		7.65	13.70		21.35	29.50
2350	Rowlock stretcher (4.50/S.F.)		310	.129		2.59	4.42		7.01	9.60
2400	Soldier course (6.75/S.F.)		200	.200		3.85	6.85		10.70	14.70
2450	Sailor course (4.50/S.F.)		290	.138		2.59	4.73		7.32	10.05
2600	Buff or gray face, running bond, (6.75/S.F.)		220	.182		4.08	6.25		10.33	14
2700	Glazed face brick, running bond	↓	210	.190	↓	9.70	6.55		16.25	20.50

Figure 9.46

Credit: *Means Building Construction Cost Data 2007*

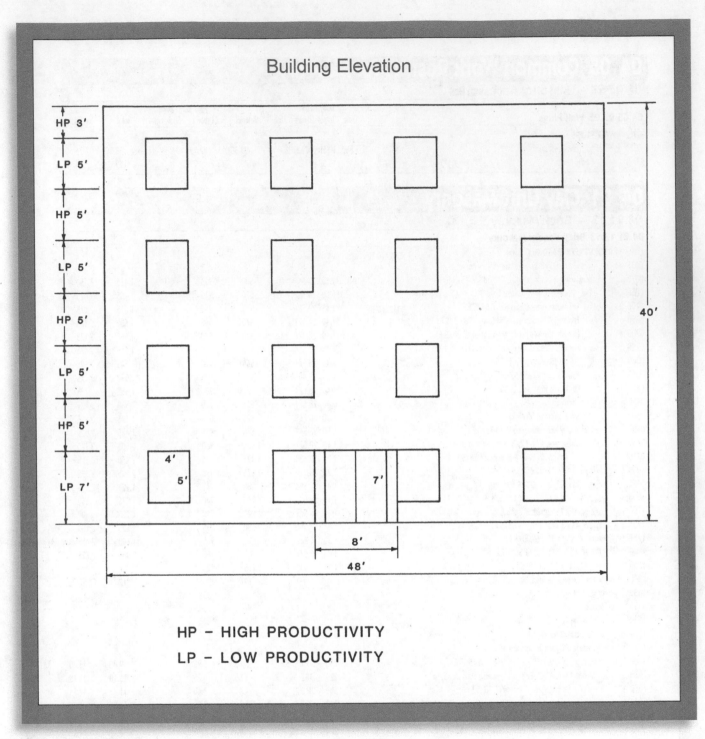

Building Elevation

HP – HIGH PRODUCTIVITY
LP – LOW PRODUCTIVITY

Figure 9.47

Brick Productivity Table

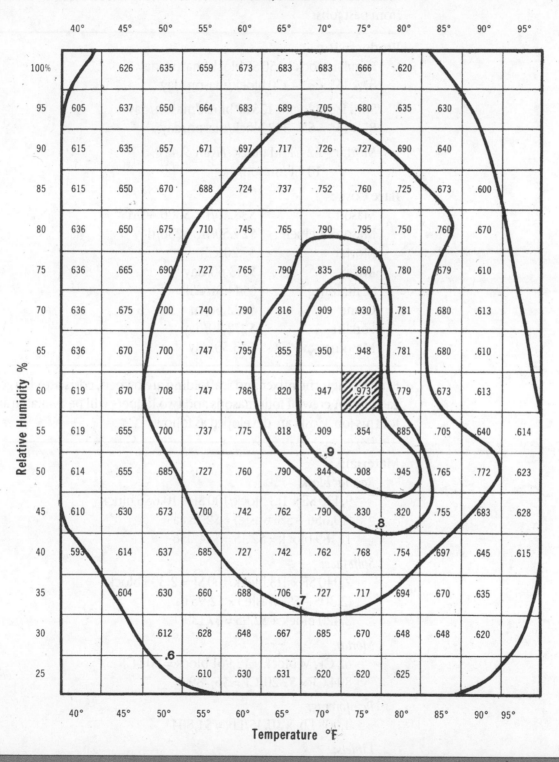

Figure 9.48

For the masonry portion of the sample building project, assume that good historical records are available. The following data have been developed from past jobs:

Productivity:

    Regular block – 150 SF block/mason/day
    8" × 16" × 8" – 170 block/mason/day

    Split face block – 125 SF block/mason/day
    8" × 16" × 6" – 142 block/mason/day

    Mortar – 62 CF/M block *(Figure 9.42)*
           10 CF/mason/day

Bare Costs:

Mason	$38.30/hr, $306.40/day
Mason's helper	$29.50/hr, $236/day
Mortar	$7.20/CF
Mixer	$113.60/day
Scaffolding	$144.51/CSF
Crane	$123.60/hr
Forklift	$272.40/day
Masonry saw	$43.50/day

Even though the mixer and scaffolding are owned, costs are charged to each job. A crew of four masons and two helpers will be available and used for this job. Costs are developed as follows:

Material:

Regular block:

    9,430 SF × 113 block/100 SF = 10,656 block
    (10,656 + 5% waste) × $2.15 ea.
    11,189 block × $2.15 = $24,056

Split face:

    2,040 SF × 113 block/100 SF = 2,305 block
    (2,305 + 5% waste) × $2.95 ea.
    2,420 block × $2.95 = $7,139

Mortar:

    62 CF/M block × 12,961 block = 804 CF
    804 CF × $7.20/CF = $5,789

Reinforcing:

    4,008 Lb. × $0.45/Lb. = $1,804

Lintels:

    Labor costs included below.
    Material costs included in Division 5.

Total Material Costs:

Regular Block	$24,056
Split Face Block	7,139
Mortar	5,789
Reinforcing	1,804
	$38,788

Labor:

Regular block:

$$\frac{9,430 \text{ SF}}{150 \text{ SF/mason/day} \times 4 \text{ masons}} = 16 \text{ days}$$

Split block:

$$\frac{2,040 \text{ SF}}{125 \text{ SF/mason/day} \times 4 \text{ masons}} = 4 \text{ days}$$

4 masons × $306.40/day × 20 days = $24,512
2 helpers × $236/day × 20 days = $9,440
1 helper × $ 44.25/hour × 20 hours = $885 (overtime)

Total Labor Costs:

Masons	$24,512
Helpers	10,325
	$34,837

Equipment:

Mixer:

$113.60/day × 20 days = $2,272

Scaffolding:

24 sections for 1 month @ $144.50/CSF × 10 CSF = $1,445

Crane:

4 hours × $123.60/hour = $494

Forklift:

10 days × $272.40/day = $2,724

Masonry saw:

10 days × $43.50/day = $435

Total Equipment Costs:

Mixer	$2,272
Scaffolding	1,445
Crane	494
Forklift	2,724
Masonry saw	431
	$7,366

The costs are entered as shown in Figure 9.49. Unit costs for materials should be obtained from or verified by local suppliers. The most accurate prices are those developed from recent data. A thorough and detailed cost control and accounting system is important for the development of such costs. Overtime costs for one helper are for mixing mortar before the start of the workday.

Sample Estimate: Division 4

CONSOLIDATED ESTIMATE

PROJECT: Office Building
LOCATION:
TAKE OFF BY: ABC
CLASSIFICATION: Division 4
ARCHITECT:
QUANTITIES BY: ABC
PRICES BY: ABC
EXTENSIONS BY: As Shown
ESTIMATE NO:
DATE: Jan-07
CHECKED BY: GHI
DEF

DESCRIPTION	SOURCE	QUANT	UNIT	MATERIAL COST	MATERIAL TOTAL	LABOR COST	LABOR TOTAL	EQUIPMENT COST	EQUIPMENT TOTAL	SUBCONTRACT COST	SUBCONTRACT TOTAL	TOTAL
Division 4: Masonry												
Regular Block 8" x 16" x 8"	04 22 10.14 1150	11189	Ea.	2.15	24056							
(Material only)												
Split Face Block 8" x 16" x 8"	04 22 10.23 6150	2420	Ea.	2.95	7139							
(Material only)												
Mortar	04 05 13.30 2100	804	CF	7.20	5789							
(Material only)												
Reinforcing	04 05 19.26 0015	4008	Lb.	0.45	1804							
(Material only)												
Labor: Masons ($38.30/hour)		80	Day			306.40	24512					
Mason Helper ($28.75/hour)		40	Day			230.00	9200					
Overtime ($43.15/hour)		20	Hour			43.15	863					
Equipment: *Mixer	01 54 33.10 1800	20	Day					113.60	2272			
Scaffolding (24 Sections)	01 54 23.70 0090	10	C.S.F.			110.00	1100	144.5	1445			
*Crane	01 54 23.70 0906	10	C.S.F.	34.50	345			123.60	494			
*Forklift	01 54 33.60 2500	4	Hour					272.40	2724			
*Masonry Saw	01 54 33.40 2040	10	Day					43.05	431			
Note: "*" in Reference Section	01 54 33.40 6000	10	Day									
Division 4 Totals				$	39,133	$	35,675	$	7,366			

Division 5: Metals

The metals portion of a building project, and the corresponding estimate, should be broken down into basic components: structural metals, metal joists, metal decks, miscellaneous metals, ornamental metals, expansion control, and fasteners. The items in a building project are shown on many different sheets of the drawings and may or may not be thoroughly listed in the specifications. This is especially true of the miscellaneous metals that are listed under Division 5. A complete and detailed review of the construction documents is therefore necessary, noting all items and requirements. Most structural steel work is subcontracted to specialty fabricators and erectors. However, the estimator for the general contractor may perform a takeoff to ensure that all specific work is included. Pricing based on total weight (tonnage) can be used to verify subcontractor prices.

Structural Metals

The various structural members should be identified and separately listed by type and size. For example:

	Size	Length	Quantity
Columns:	W8 x 67	12'-6"	16
	W12 x 58	12'-6"	8
Beams:	W14 x 30	30'-0"	10
	W14 x 30	28'-0"	6
	W14 x 26	30'-0"	16
	W24 x 55	30'-0"	12

The quantities may be converted to weight (tonnage) for pricing. This figure is based on weight per linear foot. Base plates, leveling plates, anchor bolts, and other accessories should be taken off and listed at this time. Connections should also be noted. Costs for structural steel in *Means Building Construction Cost Data* (Figures 9.50 and 9.51) are given for both individual members and for complete projects including bolted connections. If specified, costs for high strength steel and/or high strength bolts must be added separately. Welded connections should be listed by fillet size and length and priced separately. Light-gauge metal and special framing for items, such as hanging lintels, fascias, and parapets, should be taken off and priced separately. The amount of this type of metal work will vary depending on design and project requirements.

For development of a budget cost or verification of a subcontractor bid, the estimator can apply average allowance percentages to gross tonnage for:

Base plates	2 to 3%
Column splices &	
beam connections	8 to 10%
Total allowance	10 to 13% of main members

QUANTITY SHEET

PROJECT: Office Building
LOCATION:
TAKE OFF BY:

Division 5
ARCHITECT:
EXTENSIONS BY:

SHEET NO. 2 of 2
ESTIMATE NO:
DATE:
CHECKED BY:

DESCRIPTION	NO.	L	W	Lbs.	UNIT	Lbs.	UNIT	Total Lbs.	UNIT	Acc.	UNIT
Beams - Roof											
W16 x 26	4	34'				3536	Lbs.				
	8	35'				7280					
W21 x 44	7	35'				10780					
W18 x 24	1	26'				1040					
W21 x 49	4	29'				5684					
	2	30'				2940					
W21 x 62	10	29'				17980					
W27 x 94	3	30'				5580					
	2	30'				5640					
W12 x 19	1	9'				171	Lbs.	60631	Lbs.		
3/4" Shear Studs		4 3/16"								536	Ea.
Beams - Penthouse											
W21 x 49	2	30'				2940	Lbs.				
W21 x 62	1	30'				1860	Lbs.				
W16 x 26	4	35'				3640	Lbs.	8440	Lbs.		
Joists - Roof & Penthouse											
2447 (R)	30	34'	11.5			11730	Lbs.				
(R)	52	35'	11.5			20930	Lbs.				
(P)	10	30'	11.5			3450	Lbs.	36110	Lbs.		
Bottom Chord Ext.	32									32	Ea.
Bridging (R)	36	88'	0.8			2534	Lbs.				
(R)	24	29'	0.8			557	Lbs.				
(P)	12	30'	0.8			288	Lbs.	3379	Lbs.		
Deck (22 ga.)											
3" - 2nd & 3rd Floor	2	210'	90'		39900 SF						
Incl. Penthouse		70'	30'								
1 1/2" - R & P		210'	90'		18900 SF						
Edge Form	3	210'	90' + Opening							2160	LF

Figure 9.55

Sample Estimate: Division 5

CONSOLIDATED ESTIMATE

PROJECT:	Office Building
LOCATION:	
TAKE OFF BY: ABC	QUANTITIES BY: ABC
ARCHITECT:	PRICES BY:
CLASSIFICATION:	Division 5
EXTENSIONS BY:	

ESTIMATE NO:
DATE: Jan-07
CHECKED BY: GHI

DESCRIPTION	SOURCE	QUANT	UNIT	MATERIAL COST	MATERIAL TOTAL	LABOR COST	LABOR TOTAL	EQUIPMENT COST	EQUIPMENT TOTAL	SUBCONTRACT COST	SUBCONTRACT TOTAL	TOTAL COST	TOTAL
Division 5: Metals													
Structural Steel													
Columns		53,885	Lbs.										
2nd & 3rd Floor Beams		190,554											
Roof		60,631											
Penthouse		8440											
		313,510	Lbs.										
10% Connections		31,351											
		344,861	Lbs.										
Base Plate		5,275											
		350,136	Lbs.										
Total	05 12 23.77 0800	175	Ton							3198	559650		
High Strength Bolts 20 Each per Ton	05 05 23.25 0300	3500	Ea.							7.33	25655		
Anchor Bolts (material Only)	03 15 05.02 0500	112	Ea.							3.95	442		
Subtotals											$ 585,747		

Figure 9.56

Sample Estimate: Division 5

CONSOLIDATED ESTIMATE

PROJECT: Office Building	CLASSIFICATION: Division 5
LOCATION:	ESTIMATE NO:
	ARCHITECT:
TAKE OFF BY: ABC	QUANTITIES BY: ABC

PRICES BY: DEF — EXTENSIONS BY: DEF — DATE: Jan-07 — CHECKED BY: GHI

DESCRIPTION	SOURCE	QUANT	UNIT	MATERIAL COST	MATERIAL TOTAL	LABOR COST	LABOR TOTAL	EQUIPMENT COST	EQUIPMENT TOTAL	SUBCONTRACT COST	SUBCONTRACT TOTAL	TOTAL COST	TOTAL TOTAL
Division 5: (Cont'd)													
Metal Joists & Decks													
Joists - Roof and Penthouse		36,110	Lbs.										
Bridging		3,379											
		39,489	Lbs.										
Total	05 12 23.77 0800	19.75	Ton							3198	63161		
Bottom Chord Extensions	05 21 19.10 6300	32	Ea.							35.70	1142		
Composite													
Deck - 3", 22 gauge	05 31 13.50 5200	40570	S.F.							2.36	95745		
Roof - 1-1/2", 22 gauge	05 31 23.50 2400	18900	S.F.							1.78	33642		
Edge Form	05 31 33.50 7100	2160	L.F.							5.79	12506		
Shear Connector	05 05 23.85 0300	3172	Ea.							2.27	7200		
3/4" diameter x 4-3/16" long													
Subtotals													$ 213,397

Figure 9.57

To download this and other forms in this book, visit www.rsmeans.com/supplement/67303B.asp

Sample Estimate: Division 5

CONSOLIDATED ESTIMATE

PROJECT:	Office Building
LOCATION:	
TAKE OFF BY: ABC	QUANTITIES BY: ABC
CLASSIFICATION:	Division 5
ARCHITECT:	
PRICES BY:	As Shown
EXTENSIONS BY:	
SHEET NO. 3 of 3	
ESTIMATE NO:	
DATE:	Jan-07
CHECKED BY: GHI	

DESCRIPTION	SOURCE			QUANT	UNIT	MATERIAL COST	MATERIAL TOTAL	LABOR COST	LABOR TOTAL	EQUIPMENT COST	EQUIPMENT TOTAL	SUBCONTRACT COST	SUBCONTRACT TOTAL	TOTAL COST	TOTAL
Division 5: (Cont'd)															
Miscellaneous Metals															
Stairs 3'-6" w/ Rails	05 51	13.50	0200	188	Riser							527.00	99076		
Landings 10'-8" x 4'	05 51	13.50	1500	320	S.F.							76.40	24448		
Exterior Aluminum Rails	05 52	13.50	0020	150	L.F.							43.30	6495		
Lintels (Material Only)	05 12	23.45	2100	34	Ea.							23.05	784		
Subtotals												$	130,803		
Sheet 1 Subtotals												$	585,747		
Sheet 2 Subtotals												$	213,397		
Sheet 3 Subtotals												$	130,803		
Division 5 Totals												$	929,947		

Figure 9.58

Metals — R0505 Common Work Results for Metals

R050516-30 Coating Structural Steel

On field-welded jobs, the shop-applied primer coat is necessarily omitted. All painting must be done in the field and usually consists of red oxide rust inhibitive paint or an aluminum paint. The table below shows paint coverage and daily production for field painting.

See Division 09 97 13.23 for hot-dipped galvanizing and for field-applied cold galvanizing and other paints and protective coatings.

See Division 05 01 10.51 for steel surface preparation treatments such as wire brushing, pressure washing and sand blasting.

Type Construction	Surface Area per Ton	Coat	One Gallon Covers		In 8 Hrs. Person Covers		Average per Ton Spray	
			Brush	Spray	Brush	Spray	Gallons	Labor-hours
Light Structural	300 S.F. to 500 S.F.	1st	500 S.F.	455 S.F.	640 S.F.	2000 S.F.	0.9 gals.	1.6 L.H.
		2nd	450	410	800	2400	1.0	1.3
		3rd	450	410	960	3200	1.0	1.0
Medium	150 S.F. to 300 S.F.	All	400	365	1600	3200	0.6	0.6
Heavy Structural	50 S.F. to 150 S.F.	1st	400	365	1920	4000	0.2	0.2
		2nd	400	365	2000	4000	0.2	0.2
		3rd	400	365	2000	4000	0.2	0.2
Weighted Average	225 S.F.	All	400	365	1350	3000	0.6	0.6

R050521-20 Welded Structural Steel

Usual weight reductions with welded design run 10% to 20% compared with bolted or riveted connections. This amounts to about the same total cost compared with bolted structures since field welding is more expensive than bolts. For normal spans of 18' to 24' figure 6 to 7 connections per ton.

Trusses — For welded trusses add 4% to weight of main members for connections. Up to 15% less steel can be expected in a welded truss compared to one that is shop bolted. Cost of erection is the same whether shop bolted or welded.

General — Typical electrodes for structural steel welding are E6010, E6011, E60T and E70T. Typical buildings vary between 2# to 8# of weld rod per

ton of steel. Buildings utilizing continuous design require about three times as much welding as conventional welded structures. In estimating field erection by welding, it is best to use the average linear feet of weld per ton to arrive at the welding cost per ton. The type, size and position of the weld will have a direct bearing on the cost per linear foot. A typical field welder will deposit 1.8# to 2# of weld rod per hour manually. Using semiautomatic methods can increase production by as much as 50% to 75%.

R050523-10 High Strength Bolts

Common bolts (A307) are usually used in secondary connections (see Division 05 05 23.10).

High strength bolts (A325 and A490) are usually specified for primary connections such as column splices, beam and girder connections to columns, column bracing, connections for supports of operating equipment or of other live loads which produce impact or reversal of stress, and in structures carrying cranes of over 5-ton capacity.

Allow 20 field bolts per ton of steel for a 6 story office building, apartment house or light industrial building. For 6 to 12 stories allow 25 bolts per ton, and above 12 stories, 30 bolts per ton. On power stations, 20 to 25 bolts per ton are needed.

Figure 9.59

Division 6: Wood, Plastics, & Composites

Wood frame construction is still dominant in the residential construction industry in the United States. However, its use for large-scale commercial buildings has declined. There are many reasons for this trend; among them are design criteria, cost, and in some localities, building and fire code restrictions. Nevertheless, the use of wood framing for smaller suburban office buildings is still common.

Material prices for lumber fluctuate more and with greater frequency than any other building material. For this reason, when the material list is complete, it is important to obtain current, local prices for the lumber. Installation costs depend on productivity. For Division 6, accurate cost records from past jobs can be most helpful. The estimate can be tailored to the productivity of specific crews, as shown in Division 4 of the Sample Estimate.

Carpentry work can be broken down into the following categories: rough carpentry, finish carpentry and millwork, and laminated framing and decking. The rough carpentry materials can be sticks of lumber or sheets of plywood—job-site fabricated and installed, or may consist of trusses and truss joists and panelized roof systems—prefabricated, delivered, and erected by a specialty subcontractor.

Rough Carpentry

Lumber is usually estimated in board feet and purchased in 1,000 board foot quantities. A board foot is 1" x 12" x 12" (nominal) or 3/4" x 11-1/2" x 12" milled (actual). To determine board feet of a piece of framing, the nominal dimensions can be multiplied, and the result divided by 12. The final result represents the number of board feet per linear foot of that framing size.

Example: 2 x 10 joists
$2 \times 10 = 20$

$$\frac{20}{12} = 1.67 \qquad \frac{\text{Board feet}}{\text{Linear foot}}$$

The Quantity Sheet should indicate species, grade, and any type of wood preservative or fire-retardant treatment specified or required by code.

Sills, Posts, and Girders used in subfloor framing should be taken off by length and quantity. The standard lengths available vary by species and dimensions. Cut-offs are often used for blocking. Careful selection of lengths will decrease the waste factor required.

Floor Joists, shown or specified by size and spacing, should be taken off by nominal length and the quantity required. Add for double joists under partitions, headers and cripple joists at openings, overhangs, laps at bearings, and blocking or bridging.

Ceiling Joists, similar to floor joists, carry roof loads and/or ceiling finishes. Soffits and suspended ceilings should be noted and taken off separately. Ledgers may be a part of the ceiling joist system. In a flat roof system, the rafters are called joists and are usually shown as a ceiling system.

Studs required are noted on the drawings by spacing, usually 16" OC or 24" OC, with the stud size given. The linear feet of partitions with the same stud size, height, and spacing, and divided by the spacing, will give the estimator the approximate number of studs required. Additional studs for openings, corners, double top plates, sole plates, and intersecting partitions must be taken off separately. An allowance for waste should be included (or heights should be recorded as a standard length). A rule of thumb is to allow one stud for every linear foot of wall for 16" OC.

Number and Size of Openings are important. Even though there are no studs in these areas, the estimator must take off headers, subsills, king studs, trimmers, cripples, and knee studs. Where bracing and fire blocking are noted, indicate the type and quantity.

Roof Rafters vary with different types of roofs. A hip and valley, because of its complexity, has a greater material waste factor than most other roof types. Although straight gable, gambrel, and mansard roofs are not as complicated, care should be taken to ensure a good material takeoff. Roof pitches, overhangs, and soffit framing all affect the quantity of material and therefore, the costs. Rafters must be measured along the slope, not the horizontal. *(See Figure 9.64.)*

Roof Trusses are usually furnished and delivered to the job site by the truss fabricator. The high cost of job site labor and new gang nailing technology have created a small boom in truss manufacturing. Many architects' designs of wood frame and masonry bearing wall structures now include wood trusses of both the trussed rafter type and the flat chord type (also used for floors). Depending on the size of truss, hoisting equipment may be needed for erection. The estimator should obtain prices and weights from the fabricator and should determine whether or not erection is included in the fabricator's cost. Architecturally exposed trusses are typically more expensive to fabricate and erect.

Tongue and Groove Roof Decks of various woods, solid planks, or laminated construction are nominally 2" to 4" thick and are often used with glued-laminated beams or heavy timber framing. The square foot method is used to determine quantities, and consideration is given to roof pitches and non-modular areas for the amount of waste involved. The materials are purchased by board foot measurement. The conversion from square foot to board foot must allow for net sizes as opposed to board measure. In this way, loss of coverage due to the available tongue and mill lengths can be taken into account.

Sheathing on walls can be plywood of different grades and thicknesses, wallboard, or solid boards nailed directly to the studs. Insulating sheets with air infiltration barriers are often used as sheathing in colder climates. Plywood can be applied with the grain vertical, horizontal, or rarely, diagonal to the studding. Solid boards are usually nailed diagonally, but can be applied horizontally when lateral forces are not present. For solid board sheathing, add 15% to 20% more material to the takeoff when using tongue and groove, as opposed to square edge sheathing. Wallboard can be installed either horizontally or vertically, depending on wall height and fire code restrictions.

When estimating quantities of plywood or wallboard sheathing, calculate the number of sheets required by measuring the square feet of area to be covered and then dividing by sheet size. Applying these materials diagonally or on non-modular areas will create waste. This waste factor must be included in the estimate. For diagonal application of boards, plywood, or wallboard, include an additional 10% to 15% material waste factor.

Subfloors can be CDX-type plywood (with the thickness dependent on the load and span), solid boards laid diagonally or perpendicular to the joists, or tongue and groove planks. The quantity takeoff for subfloors is similar to sheathing (noted above).

Stressed Skin Plywood includes prefabricated roof panels, with or without bottom skin or tie rods, and folded plate roof panels with intermediate rafters. Stressed skin panels are typically custom-prefabricated. Takeoff is by the square foot or panel.

Structural Joists are prefabricated "beams" with wood flanges and plywood or tubular steel webs. This type of joist is spaced in accordance with the load and requires bridging and blocking supplied by the fabricator. Quantity takeoff should include the following: type, number required, length, spacing, end-bearing conditions, number of rows, length of bridging, and blocking.

Grounds are normally 1" x 2" wood strips used for case work or plaster; the quantities are estimated in LF.

Furring (1" x 2" or 3") wood strips are fastened to wood masonry or concrete walls so that wall coverings may be attached thereto. Furring may also be used on the underside of ceiling joists to fasten ceiling finishes. Quantities are estimated by LF.

Lumber and Plywood Treatments can sometimes double the costs for material. The plans and specifications should be carefully checked for required treatments— against insects, fire, or decay—as well as grade, species, and drying specifications.

An alternative method to pricing rough carpentry by the piece or linear foot is to determine quantities (in board feet) based on square feet of surface area. Appendix A of this book contains charts and tables that may be used for this second method. Also included are quantities of nails required for each type and spacing of rough framing. A rule of thumb for this method can be used to determine linear feet of framing members (such as studs, joists) based on square feet of surface area (wall, floor, ceiling):

Spacing of Framing Members	Board Feet per Square Foot Surface
12" OC	1.2 BF/SF
16" OC	1.0 BF/SF
24" OC	0.8 BF/SF

The requirements for rough carpentry, especially those for temporary construction, may not all be directly stated in the plans and specifications. These additional items may include blocking, temporary stairs, wood inserts for metal pan stairs, and railings, along with various other requirements for different trades. Temporary construction may also be included in Division 1 of the General Requirements.

Finish Carpentry & Millwork

Finish carpentry and millwork—wood rails, paneling, shelves, casements and cabinetry—are common in buildings that have no other wood. After examining the plans and specifications, it's important to determine which items will be built on-site, and which will be fabricated off-site by a millwork subcontractor. Shop drawings are often required for architectural woodwork and are usually included in the subcontract price.

Window and Door Trim may be taken off and priced by the "set" or by the linear foot. Check for jamb extensions at exterior walls. The common use of pre-hung doors and windows makes it convenient to take off this trim with the doors and windows. Exterior trim, other than door and window trim, should be taken off with the siding, since the details and dimensions are interrelated.

Paneling is taken off by the type, finish, and square foot (converted to full sheets). Be sure to list any millwork that would show up on the details. Panel siding and associated trim are taken off by the square foot and linear foot, respectively. Be sure to provide an allowance for waste.

Decorative Beams and Columns that are non-structural should be estimated separately. Decorative trim may be used to wrap exposed structural elements. Particular attention should be paid to the joinery. Long, precise joints are difficult to construct in the field.

Cabinets, Counters, and Shelves are most often priced by the linear foot or by the unit. Job-fabricated, prefabricated, and subcontracted work should be estimated separately.

Stairs should be estimated by individual component unless accurate, complete system costs have been developed from previous projects. Typical components and units for estimating are shown in Figure 9.60.

A general rule for budgeting millwork is that total costs will be two to three times the cost of the materials. Millwork is often ordered and purchased directly by the owner; when installation is the responsibility of the contractor, costs for handling, storage, and protection should be included.

Laminated Construction

Laminated construction should be listed separately, as it is frequently supplied by a specialty subcontractor. Sometimes the beams are supplied and erected by one subcontractor, and the decking installed by the general contractor or another subcontractor. The takeoff units must be adapted to the system: square foot—floor, linear foot—members, or board foot—lumber. Since the members are factory-fabricated, the plans and specifications must be submitted to a fabricator for takeoff and pricing.

06 25 Prefinished Paneling

06 25 26 – Panel Systems

06 25 26.10 Panel Systems	Crew	Daily Output	Labor-Hours	Unit	Material	2007 Bare Costs Labor	Equipment	Total	Total Incl O&P
0610 Rustic paneling, 5/8" MDF, standard, maple veneer, unfinished	2 Carp	300	.053	S.F.	21	1.96		22.96	26
5000 For prefinished paneling, see division 06 25 16.10 & 06 25 13.10									

06 26 Board Paneling

06 26 13 – Profile Board Paneling

06 26 13.10 Paneling, Boards

		Crew	Daily Output	Labor-Hours	Unit	Material	2007 Bare Costs Labor	Equipment	Total	Total Incl O&P
0010	PANELING, BOARDS									
6400	Wood board paneling, 3/4" thick, knotty pine	2 Carp	300	.053	S.F.	1.41	1.96		3.37	4.60
6500	Rough sawn cedar		300	.053		1.81	1.96		3.77	5.05
6700	Redwood, clear, 1" x 4" boards		300	.053		4.22	1.96		6.18	7.70
6900	Aromatic cedar, closet lining, boards	↓	275	.058	↓	3.25	2.14		5.39	6.90

06 43 Wood Stairs and Railings

06 43 13 – Wood Stairs

06 43 13.20 Prefabricated Wood Stairs

		Crew	Daily Output	Labor-Hours	Unit	Material	2007 Bare Costs Labor	Equipment	Total	Total Incl O&P
0010	PREFABRICATED WOOD STAIRS									
0100	Box stairs, prefabricated, 3'-0" wide									
0110	Oak treads, up to 14 risers	2 Carp	39	.410	Riser	75.50	15.05		90.55	107
0600	With pine treads for carpet, up to 14 risers	"	39	.410	"	49	15.05		64.05	77.50
1100	For 4' wide stairs, add				Flight	25%				
1550	Stairs, prefabricated stair handrail with balusters	1 Carp	30	.267	L.F.	65	9.80		74.80	87
1700	Basement stairs, prefabricated, pine treads									
1710	Pine risers, 3' wide, up to 14 risers	2 Carp	52	.308	Riser	49	11.30		60.30	71.50
4000	Residential, wood, oak treads, prefabricated		1.50	10.667	Flight	985	390		1,375	1,675
4200	Built in place	↓	.44	36.364	"	1,475	1,325		2,800	3,700
4400	Spiral, oak, 4'-6" diameter, unfinished, prefabricated,									
4500	incl. railing, 9' high	2 Carp	1.50	10.667	Flight	4,675	390		5,065	5,725

06 43 13.30 Wood Stair Components

		Crew	Daily Output	Labor-Hours	Unit	Material	2007 Bare Costs Labor	Equipment	Total	Total Incl O&P
0010	WOOD STAIR COMPONENTS									
0020	Balusters, turned, 3" high, pine, minimum	1 Carp	28	.286	Ea.	3.77	10.50		14.27	20.50
0100	Maximum		26	.308		19.40	11.30		30.70	39
0300	30" high birch balusters, minimum		28	.286		6.85	10.50		17.35	24
0400	Maximum		26	.308		27.50	11.30		38.80	47.50
0600	42" high, pine balusters, minimum		27	.296		5	10.85		15.85	22.50
0700	Maximum		25	.320		28	11.75		39.75	49
0900	42" high birch balusters, minimum		27	.296		10.90	10.85		21.75	29
1000	Maximum		25	.320	↓	39.50	11.75		51.25	61.50
1050	Baluster, stock pine, 1-1/4" x 1-1/4"		240	.033	L.F.	3.18	1.22		4.40	5.40
1100	1-3/4" x 1-3/4"		220	.036	"	9.10	1.33		10.43	12.10
1200	Newels, 3-1/4" wide, starting, minimum		7	1.143	Ea.	38	42		80	108
1300	Maximum		6	1.333		325	49		374	435
1500	Landing, minimum		5	1.600		106	58.50		164.50	209
1600	Maximum		4	2	↓	355	73.50		428.50	510
1800	Railings, oak, built-up, minimum		60	.133	L.F.	32.50	4.89		37.39	43
1900	Maximum		55	.145		47	5.35		52.35	60
2100	Add for sub rail		110	.073		5.40	2.67		8.07	10.10
2300	Risers, beech, 3/4" x 7-1/2" high		64	.125		5.90	4.59		10.49	13.65
2400	Fir, 3/4" x 7-1/2" high		64	.125		1.63	4.59		6.22	8.95
2600	Oak, 3/4" x 7-1/2" high	↓	64	.125	↓	6.20	4.59		10.79	14

Figure 9.60

Credit: *Means Building Construction Cost Data 2007*

Sample Estimate: Division 6

The estimate for Division 6—Wood, Plastics, & Composites—of the sample project is minimal and straightforward. The estimate sheet is shown in Figure 9.61. The only item included for rough carpentry is fire-retardant treated blocking. Blocking requirements may or may not be shown on the drawings or stated in the specifications, but in this case, previous experience dictates that an allowance be included. The only other carpentry for the project involves wall paneling at the elevator lobby of each floor and vanities in the women's restrooms. For custom quality finish work, it is recommended that the craftsman performing the work be consulted. This person will be able to estimate the required time based on experience.

Sample Estimate: Division 6

CONSOLIDATED ESTIMATE

PROJECT: Office Building   CLASSIFICATION: Division 6   ESTIMATE NO:
LOCATION: ARCHITECT: DATE: Jan-07
TAKE OFF BY: ABC   QUANTITIES BY: ABC   PRICES BY: ABC   CHECKED BY: GHI

EXTENSIONS BY: PRICES BY: DEF

DESCRIPTION	SOURCE	QUANT	UNIT	MATERIAL COST	MATERIAL TOTAL	LABOR COST	LABOR TOTAL	EQUIPMENT COST	EQUIPMENT TOTAL	SUBCONTRACT COST	SUBCONTRACT TOTAL	TOTAL COST	TOTAL
Division 6: Wood, Plastics and Composite													
Rough Carpentry													
Blocking	06 11 10.02 2740	0.1	MBF	560.00	56	2100	210						
Fire Treatment	06 05 73.10 0400	0.1	MBF	330.00	33								
Finish Carpentry													
Paneling@ Elevator Lobby	06 25 16.10 2600	1800	S.F.	2.41	4338	1.47	2646						
Division 6 Totals					$ 4,427		$ 2,856						

Figure 9.61

215

To download this and other forms in this book, visit www.rsmeans.com/supplement/67303B.asp

Division 7: Thermal & Moisture Protection

This division includes materials for sealing the outside of a building—for protection against moisture and air infiltration—as well as insulation and associated accessories. When reviewing the plans and specifications, the estimator should visualize the construction process, and thus determine all probable areas where these materials will be found in or on a building. The technique used for quantity takeoff depends on the specific materials and installation methods.

Waterproofing

Waterproofing is made up of the following components:

- Dampproofing
- Vapor barriers
- Caulking and sealants
- Sheet and membrane
- Integral cement coatings

A distinction should be made between dampproofing and waterproofing. Dampproofing is used to inhibit the migration of moisture or water vapor. In most cases, dampproofing will not stop the flow of water (even at minimal pressures). Waterproofing, on the other hand, consists of a continuous, impermeable membrane, and is used to prevent or stop the flow of water.

Dampproofing usually consists of one or two bituminous coatings applied to foundation walls from about the finished grade line to the bottom of the footings. The areas involved are calculated from the total height of the dampproofing and the length of the wall. After separate areas are figured and added together to provide a total square foot area, a unit cost per square foot can be selected for the type of material, the number of coats, and the method of application specified for the building.

Waterproofing at or below grade with elastomeric sheets or membranes is estimated on the same basis as dampproofing, with two basic exceptions. First, the installed unit costs for the elastomeric sheets do not include bonding adhesive or splicing tape, which must be figured as an additional cost. Second, the membrane waterproofing under slabs must be estimated separately from the higher-cost installation on walls. In all cases, unit costs are per square foot of covered surface.

For walls below grade, protection board is often specified to prevent damage to the barrier when the excavation is backfilled. Rigid foam insulation installed outside of the barrier may also serve a protective function. Metallic coating material may be applied to floors or walls, usually on the interior or dry side, after the masonry surface has been prepared (usually by chipping) for bonding to the new material. The unit cost per square foot for these materials depends on the thickness of the material, the position of the area to be covered, and the preparation required. In many cases, these materials must be applied in locations where access is difficult and under the control of others. The estimator should make an allowance for delays caused by this problem.

Caulking and sealants are usually applied on the exterior of the building, except for certain special conditions on the interior. In most cases, caulking and sealing are done to prevent water and/or air from entering a building, and are usually specified at joints, expansion joints, control joints, door and window frames, and in places where dissimilar materials meet over the surface of the building exterior. To estimate the installed cost of this type of material, two things must be determined. First, the estimator must note (from the specifications) the kind of material to be used for each caulking or sealing job. Second, the dimensions of the joints to be caulked or sealed must be measured on the plans, with attention given to any requirements for backer rods. With this information, it is then possible to select the applicable cost per linear foot and multiply it by the total length in feet. The result is an estimated cost for each kind of caulking or sealing on the job. Caulking and sealing may often be overlooked as incidental items. They may, in fact, represent a significant cost, depending on the type of construction.

The specifications may require testing of the integrity of installed waterproofing in certain cases. If required, adequate time must be allowed and costs included in the estimate.

Insulation

Insulation is available in various types and forms:

- Batt or roll
- Blown-in
- Board (rigid and semi-rigid)
- Cavity masonry
- Perimeter foundation
- Poured in place
- Reflective
- Roof
- Sprayed

Insulation is primarily used to reduce heat transfer through the exterior enclosure of the building. The type and form of this insulation will vary according to its location in the structure and the size of the space it occupies. Major insulation types include mineral granules, fibers and foams, vegetable fibers and solids, and plastic foams. These materials may be required around foundations, on or inside walls, and under roofing. Many different details of the drawings must be examined in order to determine types, methods, and quantities of insulation. The cost of insulation depends on the type of material, its form (loose, granular, batt or boards), its thickness in inches, the method of installation, and the total area in square feet.

07 26 Vapor Retarders
07 26 10 – Vapor Retarders

07 26 10.10 Vapor Retarders	Crew	Daily Output	Labor-Hours	Unit	Material	2007 Bare Costs Labor	Equipment	Total	Total Incl O&P
0010 **VAPOR RETARDERS**									
0020 Aluminum and kraft laminated, foil 1 side	1 Carp	37	.216	Sq.	5.10	7.95		13.05	18
0100 Foil 2 sides		37	.216		8.05	7.95		16	21
0400 Asphalt felt sheathing paper, 15#	↓	37	.216	↓	4.23	7.95		12.18	17
0450 Housewrap, exterior, spun bonded polypropylene									
0470 Small roll	1 Carp	3800	.002	S.F.	.24	.08		.32	.38
0480 Large roll	"	4000	.002	"	.13	.07		.20	.25
0500 Material only, 3' x 111.1' roll				Ea.	80			80	88
0520 9' x 111.1' roll				"	130			130	143
0600 Polyethylene vapor barrier, standard, .002" thick	1 Carp	37	.216	Sq.	1.08	7.95		9.03	13.55
0700 .004" thick		37	.216		3.03	7.95		10.98	15.70
0900 .006" thick		37	.216		4.68	7.95		12.63	17.50
1200 .010" thick		37	.216		5.35	7.95		13.30	18.20
1300 Clear reinforced, fire retardant, .008" thick		37	.216		9.75	7.95		17.70	23
1350 Cross laminated type, .003" thick		37	.216		6.85	7.95		14.80	19.85
1400 .004" thick		37	.216		7.50	7.95		15.45	20.50
1500 Red rosin paper, 5 sq rolls, 4 lb per square		37	.216		2.09	7.95		10.04	14.65
1600 5 lbs. per square		37	.216		2.80	7.95		10.75	15.45
1800 Reinf. waterproof, .002" polyethylene backing, 1 side		37	.216		5.10	7.95		13.05	17.95
1900 2 sides	↓	37	.216		6.70	7.95		14.65	19.70
2100 Roof deck vapor barrier, class 1 metal decks	1 Rofc	37	.216		14.75	6.90		21.65	28
2200 For all other decks	"	37	.216		10.50	6.90		17.40	23.50
2400 Waterproofed kraft with sisal or fiberglass fibers, minimum	1 Carp	37	.216		5.50	7.95		13.45	18.40
2500 Maximum	"	37	.216	↓	13.60	7.95		21.55	27.50

07 31 Shingles and Shakes
07 31 13 – Asphalt Shingles

07 31 13.10 Asphalt Shingles		Crew	Daily Output	Labor-Hours	Unit	Material	2007 Bare Costs Labor	Equipment	Total	Total Incl O&P
0010 **ASPHALT SHINGLES**										
0100 Standard strip shingles										
0150 Inorganic, class A, 210-235 lb/sq	CN	1 Rofc	5.50	1.455	Sq.	41.50	46.50		88	125
0155 Pneumatic nailed			7	1.143		41.50	36.50		78	108
0200 Organic, class C, 235-240 lb/sq			5	1.600		44.50	51		95.50	136
0205 Pneumatic nailed		↓	6.25	1.280	↓	44.50	40.50		85	118
0250 Standard, laminated multi-layered shingles										
0300 Class A, 240-260 lb/sq		1 Rofc	4.50	1.778	Sq.	52	56.50		108.50	154
0305 Pneumatic nailed			5.63	1.422		52	45		97	134
0350 Class C, 260-300 lb/square, 4 bundles/square			4	2		51.50	63.50		115	165
0355 Pneumatic nailed		↓	5	1.600	↓	51.50	51		102.50	144
0400 Premium, laminated multi-layered shingles										
0450 Class A, 260-300 lb, 4 bundles/sq		1 Rofc	3.50	2.286	Sq.	67.50	72.50		140	197
0455 Pneumatic nailed			4.37	1.831		67.50	58		125.50	173
0500 Class C, 300-385 lb/square, 5 bundles/square			3	2.667		75.50	85		160.50	228
0505 Pneumatic nailed			3.75	2.133		75.50	68		143.50	199
0800 #15 felt underlayment			64	.125		4.23	3.98		8.21	11.40
0825 #30 felt underlayment			58	.138		7.05	4.39		11.44	15.20
0850 Self adhering polyethylene and rubberized asphalt underlayment			22	.364	↓	49.50	11.55		61.05	74
0900 Ridge shingles			330	.024	L.F.	1.44	.77		2.21	2.89
0905 Pneumatic nailed		↓	412.50	.019	"	1.44	.62		2.06	2.63
1000 For steep roofs (7 to 12 pitch or greater), add							50%			

Figure 9.68

Credit: *Means Building Construction Cost Data 2007*

07 72 Roof Accessories

07 72 23 – Relief Vents

07 72 23.10 Roof Vents	Crew	Daily Output	Labor-Hours	Unit	Material	2007 Bare Costs Labor	2007 Bare Costs Equipment	Total	Total Incl O&P
0010 **ROOF VENTS**									
0020 Mushroom shape, for built-up roofs, aluminum	1 Rofc	30	.267	Ea.	68.50	8.50		77	90
0100 PVC, 6" high	"	30	.267	"	28.50	8.50		37	46

07 72 26 – Ridge Vents

07 72 26.10 Ridge Vents	Crew	Daily Output	Labor-Hours	Unit	Material	Labor	Equipment	Total	Total Incl O&P
0010 **RIDGE VENTS**									
0100 Aluminum strips, mill finish	1 Rofc	160	.050	L.F.	1.64	1.59		3.23	4.50
0150 Painted finish		160	.050	"	2.08	1.59		3.67	4.99
0200 Connectors		48	.167	Ea.	4.26	5.30		9.56	13.70
0300 End caps		48	.167	"	1.64	5.30		6.94	10.80
0400 Galvanized strips		160	.050	L.F.	2.14	1.59		3.73	5.05
0430 Molded polyethylene, shingles not included		160	.050	"	2.85	1.59		4.44	5.85
0440 End plugs		48	.167	Ea.	1.64	5.30		6.94	10.80
0450 Flexible roll, shingles not included		160	.050	L.F.	2.31	1.59		3.90	5.25
2300 Ridge vent strip, mill finish	1 Shee	155	.052	"	2.68	2.25		4.93	6.40

07 72 33 – Roof Hatches

07 72 33.10 Roof Hatches	Crew	Daily Output	Labor-Hours	Unit	Material	Labor	Equipment	Total	Total Incl O&P
0010 **ROOF HATCHES**									
0500 2'-6" x 3', aluminum curb and cover	G-3	10	3.200	Ea.	715	116		831	965
0520 Galvanized steel curb and aluminum cover	CN	10	3.200		695	116		811	945
0540 Galvanized steel curb and cover		10	3.200		555	116		671	790
0600 2'-6" x 4'-6", aluminum curb and cover		9	3.556		825	129		954	1,100
0800 Galvanized steel curb and aluminum cover		9	3.556		600	129		729	860
0900 Galvanized steel curb and cover		9	3.556		775	129		904	1,050
1100 4' x 4' aluminum curb and cover		8	4		1,475	145		1,620	1,825
1120 Galvanized steel curb and aluminum cover		8	4		600	145		745	885
1140 Galvanized steel curb and cover		8	4		1,425	145		1,570	1,775
1200 2'-6" x 8'-0", aluminum curb and cover		6.60	4.848		1,400	175		1,575	1,825
1400 Galvanized steel curb and aluminum cover		6.60	4.848		1,150	175		1,325	1,550
1500 Galvanized steel curb and cover		6.60	4.848		1,175	175		1,350	1,575
1800 For plexiglass panels, 2'-6" x 3'-0", add to above					400			400	440

07 72 36 – Smoke Vents

07 72 36.10 Smoke Hatches	Crew	Daily Output	Labor-Hours	Unit	Material	Labor	Equipment	Total	Total Incl O&P
0010 **SMOKE HATCHES**									
0200 For 3'-0" long, add to roof hatches from division 07 72 33.10				Ea.	25%	5%			
0250 For 4'-0" long, add to roof hatches from division 07 72 33.10					20%	5%			
0300 For 8'-0" long, add to roof hatches from division 07 72 33.10					10%	5%			

07 72 36.20 Smoke Vents	Crew	Daily Output	Labor-Hours	Unit	Material	Labor	Equipment	Total	Total Incl O&P
0010 **SMOKE VENTS**									
0100 4' x 4' aluminum cover and frame	G-3	13	2.462	Ea.	1,675	89		1,764	2,000
0200 Galvanized steel cover and frame		13	2.462		1,475	89		1,564	1,775
0300 4' x 8' aluminum cover and frame		8	4		2,300	145		2,445	2,750
0400 Galvanized steel cover and frame		8	4		1,950	145		2,095	2,350

07 72 53 – Snow Guards

07 72 53.10 Snow Guards	Crew	Daily Output	Labor-Hours	Unit	Material	Labor	Equipment	Total	Total Incl O&P
0010 **SNOW GUARDS**									
0100 Slate & asphalt shingle roofs	1 Rofc	160	.050	Ea.	9	1.59		10.59	12.60
0200 Standing seam metal roofs		48	.167		17	5.30		22.30	27.50
0300 Surface mount for metal roofs		48	.167		7.25	5.30		12.55	17
0400 Double rail pipe type, including pipe		130	.062	L.F.	18.05	1.96		20.01	23

Figure 9.69

Credit: *Means Building Construction Cost Data 2007*

07 72 Roof Accessories

07 72 73 – Pitch Pockets

07 72 73.10 Pitch Pockets

07 72 73.10 Pitch Pockets	Crew	Daily Output	Labor-Hours	Unit	Material	2007 Bare Costs Labor	Equipment	Total	Total Incl O&P
0010 **PITCH POCKETS**									
0100 Adjustable, 4" to 7", welded corners, 4" deep	1 Rofc	48	.167	Ea.	11	5.30		16.30	21
0200 Side extenders, 6"	"	240	.033	"	2.20	1.06		3.26	4.22

07 72 80 – Vents

07 72 80.30 Vents

	Crew	Daily Output	Labor-Hours	Unit	Material	2007 Bare Costs Labor	Equipment	Total	Total Incl O&P
0010 **VENTS**									
0020 Plastic, for insulated decks, 1 per M.S.F., minimum	1 Rofc	40	.200	Ea.	13	6.35		19.35	25
0100 Maximum		20	.400		30	12.70		42.70	54.50
0300 Aluminum	↓	30	.267		13	8.50		21.50	28.50
0800 Polystyrene baffles, 12" wide for 16" O.C. rafter spacing	1 Carp	90	.089		.33	3.26		3.59	5.45
0900 For 24" O.C. rafter spacing	"	110	.073	↓	.54	2.67		3.21	4.75

07 81 Applied Fireproofing

07 81 16 – Cementitious Fireproofing

07 81 16.10 Sprayed Cementitous Fireproofing

	Crew	Daily Output	Labor-Hours	Unit	Material	2007 Bare Costs Labor	Equipment	Total	Total Incl O&P
0010 **SPRAYED CEMENTITOUS FIREPROOFING**									
0050 Not incl tamping or canvas protection									
0100 1" thick, on flat plate steel	G-2	3000	.008	S.F.	.45	.24	.04	.73	.90
0200 Flat decking		2400	.010		.45	.30	.05	.80	1.01
0400 Beams		1500	.016		.45	.49	.08	1.02	1.32
0500 Corrugated or fluted decks		1250	.019		.67	.58	.10	1.35	1.74
0700 Columns, 1-1/8" thick		1100	.022		.50	.66	.11	1.27	1.68
0800 2-3/16" thick	↓	700	.034	↓	.96	1.04	.17	2.17	2.84
0850 For tamping, add						10%			
0900 For canvas protection, add	G-2	5000	.005	S.F.	.06	.15	.02	.23	.32
1000 Acoustical sprayed, 1" thick, finished, straight work, minimum		520	.046		.47	1.40	.23	2.10	2.92
1100 Maximum		200	.120		.50	3.64	.61	4.75	6.75
1300 Difficult access, minimum		225	.107		.50	3.24	.54	4.28	6.10
1400 Maximum	↓	130	.185	↓	.56	5.60	.94	7.10	10.20
1500 Intumescent epoxy fireproofing on wire mesh, 3/16" thick									
1550 1 hour rating, exterior use	G-2	136	.176	S.F.	5.85	5.35	.90	12.10	15.65
1600 Magnesium oxychloride, 35# to 40# density, 1/4" thick		3000	.008		1.23	.24	.04	1.51	1.76
1650 1/2" thick		2000	.012		2.47	.36	.06	2.89	3.35
1700 60# to 70# density, 1/4" thick		3000	.008		1.63	.24	.04	1.91	2.20
1750 1/2" thick		2000	.012		3.28	.36	.06	3.70	4.24
2000 Vermiculite cement, troweled or sprayed, 1/4" thick		3000	.008		1.11	.24	.04	1.39	1.63
2050 1/2" thick	↓	2000	.012	↓	2.21	.36	.06	2.63	3.06

Figure 9.70

Credit: *Means Building Construction Cost Data 2007*

Exercise for Application on Installing Contractor's Overhead and Profit

| Line Number | Bare Costs | | Total Including O & P |
	Material	Labor	
07 72 33.10 1200	$1,400.00	$175.00	
07 72 36.10 0300	(10%) 140.00	(5%) 8.75	
	1,540.00	183.75	
Overhead & Profit	(10%) 154.00	(54.8%) 100.70	
	$1,694.00	$284.45	$1,978.45

Figure 9.71

Crews

Crew No.	Bare Costs Hr.	Bare Costs Daily	Incl. Subs O & P Hr.	Incl. Subs O & P Daily	Cost Per Labor-Hour Bare Costs	Cost Per Labor-Hour Incl. O&P
Crew E-17	Hr.	Daily	Hr.	Daily	Bare Costs	Incl. O&P
1 Structural Steel Foreman	$43.35	$346.80	$78.50	$628.00	$42.35	$76.70
1 Structural Steel Worker	41.35	330.80	74.90	599.20		
16 L.H., Daily Totals		$677.60		$1227.20	$42.35	$76.70
Crew E-18	Hr.	Daily	Hr.	Daily	Bare Costs	Incl. O&P
1 Structural Steel Foreman	$43.35	$346.80	$78.50	$628.00	$41.16	$72.21
3 Structural Steel Workers	41.35	992.40	74.90	1797.60		
1 Equipment Operator (med.)	38.40	307.20	57.85	462.80		
1 Lattice Boom Crane, 20 Ton		976.70		1074.37	24.42	26.86
40 L.H., Daily Totals		$2623.10		$3962.77	$65.58	$99.07
Crew E-19	Hr.	Daily	Hr.	Daily	Bare Costs	Incl. O&P
1 Structural Steel Worker	$41.35	$330.80	$74.90	$599.20	$40.52	$69.65
1 Structural Steel Foreman	43.35	346.80	78.50	628.00		
1 Equip. Oper. (light)	36.85	294.80	55.55	444.40		
1 Lattice Boom Crane, 20 Ton		976.70		1074.37	40.70	44.77
24 L.H., Daily Totals		$1949.10		$2745.97	$81.21	$114.42
Crew E-20	Hr.	Daily	Hr.	Daily	Bare Costs	Incl. O&P
1 Structural Steel Foreman	$43.35	$346.80	$78.50	$628.00	$40.48	$70.51
5 Structural Steel Workers	41.35	1654.00	74.90	2996.00		
1 Equip. Oper. (crane)	39.80	318.40	60.00	480.00		
1 Oiler	33.90	271.20	51.10	408.80		
1 Lattice Boom Crane, 40 Ton		1268.00		1394.80	19.81	21.79
64 L.H., Daily Totals		$3858.40		$5907.60	$60.29	$92.31
Crew E-22	Hr.	Daily	Hr.	Daily	Bare Costs	Incl. O&P
1 Skilled Worker Foreman	$40.00	$320.00	$62.25	$498.00	$38.67	$60.18
2 Skilled Workers	38.00	608.00	59.15	946.40		
24 L.H., Daily Totals		$928.00		$1444.40	$38.67	$60.18
Crew E-24	Hr.	Daily	Hr.	Daily	Bare Costs	Incl. O&P
3 Structural Steel Workers	$41.35	$992.40	$74.90	$1797.60	$40.61	$70.64
1 Equipment Operator (medium)	38.40	307.20	57.85	462.80		
1 -25 Ton Crane		739.60		813.56	23.11	25.42
32 L.H., Daily Totals		$2039.20		$3073.96	$63.73	$96.06
Crew E-25	Hr.	Daily	Hr.	Daily	Bare Costs	Incl. O&P
1 Welder Foreman	$43.35	$346.80	$78.50	$628.00	$43.35	$78.50
1 Cutting Torch		17.00		18.70		
1 Gase		64.80		71.28	10.23	11.25
8 L.H., Daily Totals		$428.60		$717.98	$53.58	$89.75
Crew F-3	Hr.	Daily	Hr.	Daily	Bare Costs	Incl. O&P
4 Carpenters	$36.70	$1174.40	$57.15	$1828.80	$37.32	$57.72
1 Equip. Oper. (crane)	39.80	318.40	60.00	480.00		
1 Hyd. Crane, 12 Ton		724.00		796.40	18.10	19.91
40 L.H., Daily Totals		$2216.80		$3105.20	$55.42	$77.63
Crew F-4	Hr.	Daily	Hr.	Daily	Bare Costs	Incl. O&P
4 Carpenters	$36.70	$1174.40	$57.15	$1828.80	$36.75	$56.62
1 Equip. Oper. (crane)	39.80	318.40	60.00	480.00		
1 Equip. Oper. Oiler	33.90	271.20	51.10	408.80		
1 Hyd. Crane, 55 Ton		1060.00		1166.00	22.08	24.29
48 L.H., Daily Totals		$2824.00		$3883.60	$58.83	$80.91
Crew F-5	Hr.	Daily	Hr.	Daily	Bare Costs	Incl. O&P
1 Carpenter Foreman	$38.70	$309.60	$60.25	$482.00	$37.20	$57.92
3 Carpenters	36.70	880.80	57.15	1371.60		
32 L.H., Daily Totals		$1190.40		$1853.60	$37.20	$57.92

Crew No.	Bare Costs Hr.	Bare Costs Daily	Incl. Subs O & P Hr.	Incl. Subs O & P Daily	Cost Per Labor-Hour Bare Costs	Cost Per Labor-Hour Incl. O&P
Crew F-6	Hr.	Daily	Hr.	Daily	Bare Costs	Incl. O&P
2 Carpenters	$36.70	$587.20	$57.15	$914.40	$34.14	$52.76
2 Building Laborers	28.75	460.00	44.75	716.00		
1 Equip. Oper. (crane)	39.80	318.40	60.00	480.00		
1 Hyd. Crane, 12 Ton		724.00		796.40	18.10	19.91
40 L.H., Daily Totals		$2089.60		$2906.80	$52.24	$72.67
Crew F-7	Hr.	Daily	Hr.	Daily	Bare Costs	Incl. O&P
2 Carpenters	$36.70	$587.20	$57.15	$914.40	$32.73	$50.95
2 Building Laborers	28.75	460.00	44.75	716.00		
32 L.H., Daily Totals		$1047.20		$1630.40	$32.73	$50.95
Crew G-1	Hr.	Daily	Hr.	Daily	Bare Costs	Incl. O&P
1 Roofer Foreman	$33.80	$270.40	$57.30	$458.40	$29.67	$50.33
4 Roofers, Composition	31.80	1017.60	53.95	1726.40		
2 Roofer Helpers	23.35	373.60	39.60	633.60		
1 Application Equipment		153.20		168.52		
1 Tar Kettle/Pot		62.90		69.19		
1 Crew Truck		116.20		127.82	5.93	6.53
56 L.H., Daily Totals		$1993.90		$3183.93	$35.61	$56.86
Crew G-2	Hr.	Daily	Hr.	Daily	Bare Costs	Incl. O&P
1 Plasterer	$33.55	$268.40	$50.75	$406.00	$30.35	$46.33
1 Plasterer Helper	28.75	230.00	43.50	348.00		
1 Building Laborer	28.75	230.00	44.75	358.00		
1 Grout Pump, 50 C.F./hr		122.00		134.20	5.08	5.59
24 L.H., Daily Totals		$850.40		$1246.20	$35.43	$51.92
Crew G-2A	Hr.	Daily	Hr.	Daily	Bare Costs	Incl. O&P
1 Roofer, composition	$31.80	$254.40	$53.95	$431.60	$27.97	$46.10
1 Roofer Helper	23.35	186.80	39.60	316.80		
1 Building Laborer	28.75	230.00	44.75	358.00		
1 Grout Pump, 50 C.F./hr		122.00		134.20	5.08	5.59
24 L.H., Daily Totals		$793.20		$1240.60	$33.05	$51.69
Crew G-3	Hr.	Daily	Hr.	Daily	Bare Costs	Incl. O&P
2 Sheet Metal Workers	$43.55	$696.80	$67.15	$1074.40	$36.15	$55.95
2 Building Laborers	28.75	460.00	44.75	716.00		
32 L.H., Daily Totals		$1156.80		$1790.40	$36.15	$55.95
Crew G-4	Hr.	Daily	Hr.	Daily	Bare Costs	Incl. O&P
1 Labor Foreman (outside)	$30.75	$246.00	$47.90	$383.20	$29.42	$45.80
2 Building Laborers	28.75	460.00	44.75	716.00		
1 Flatbed Truck, gas, 1.5 Ton		141.40		155.54		
1 Air Compressor, 160 C.F.M.		122.60		134.86	11.00	12.10
24 L.H., Daily Totals		$970.00		$1389.60	$40.42	$57.90
Crew G-5	Hr.	Daily	Hr.	Daily	Bare Costs	Incl. O&P
1 Roofer Foreman	$33.80	$270.40	$57.30	$458.40	$28.82	$48.88
2 Roofers, Composition	31.80	508.80	53.95	863.20		
2 Roofer Helpers	23.35	373.60	39.60	633.60		
1 Application Equipment		153.20		168.52	3.83	4.21
40 L.H., Daily Totals		$1306.00		$2123.72	$32.65	$53.09
Crew G-6A	Hr.	Daily	Hr.	Daily	Bare Costs	Incl. O&P
2 Roofers, Composition	$31.80	$508.80	$53.95	$863.20	$31.80	$53.95
1 Small Compressor, Electric		9.55		10.51		
2 Pneumatic Nailers		41.30		45.43	3.18	3.50
16 L.H., Daily Totals		$559.65		$919.13	$34.98	$57.45

Figure 9.72

Credit: *Means Building Construction Cost Data 2007*

CONSOLIDATED ESTIMATE

PROJECT: Office Building	CLASSIFICATION: Division 8
LOCATION:	ARCHITECT:
TAKE OFF BY: ABC	PRICES BY: As Shown
QUANTITIES BY: ABC	EXTENSIONS BY:
	CHECKED BY: GHI

SHEET NO. 3 of 3
ESTIMATE NO:
DATE: Jan-07

SOURCE	DESCRIPTION	QUANT	UNIT	MATERIAL DEF COST	MATERIAL TOTAL	LABOR DEF COST	LABOR TOTAL	EQUIPMENT DEF COST	EQUIPMENT TOTAL	SUBCONTRACT COST	SUBCONTRACT TOTAL	TOTAL COST	TOTAL TOTAL
	Division 8: (Cont'd)												
	Window/Curtain Wall			TELEPHONE QUOTE (Including Sales Tax)									
08 44 13.10 0050		25060	S.F.								1158000		
	Subtotals												
	Sheet 1 Subtotals				$ 34,477		$ 5,822				$ 1,158,000		
	Sheet 2 Subtotals				$ 2,488		$ 396				$ 18,869		
	Sheet 3 Subtotals										$ 1,158,000		
	Division 8 Totals				$ 36,965		$ 6,218				$ 1,176,869		

Figure 9.80

To download this and other forms in this book, visit **www.rsmeans.com/supplement/67303B.asp**

08 05 Common Work Results for Openings

08 05 05 – Selective Windows and Doors Demolition

08 05 05.20 Selective Demolition of Windows	Crew	Daily Output	Labor-Hours	Unit	Material	2007 Bare Costs Labor	Equipment	Total	Total Incl O&P
5040	Average	1 Carp	4	2	Ea.		73.50		73.50	114
5080	Maximum	↓	2	4	↓		147		147	229

08 11 Metal Doors and Frames

08 11 16 – Aluminum Doors and Frames

08 11 16.10 Aluminum Doors and Frames

		Crew	Daily Output	Labor-Hours	Unit	Material	Labor	Equipment	Total	Total Incl O&P
0010	**ALUMINUM DOORS AND FRAMES**, entrance, narrow stile									
0015	Standard hardware, clear finish, not incl. glass, 2'-6" x 7'-0" opng.	2 Sswk	2	8	Ea.	810	330		1,140	1,500
0020	3'-0" x 7'-0" opening		2	8		650	330		980	1,325
0030	3'-6" x 7'-0" opening		2	8		625	330		955	1,300
0100	3'-0" x 10'-0" opening, 3' high transom		1.80	8.889		945	370		1,315	1,725
0200	3'-6" x 10'-0" opening, 3' high transom		1.80	8.889		945	370		1,315	1,725
0280	5'-0" x 7'-0" opening		2	8		1,025	330		1,355	1,725
0300	6'-0" x 7'-0" opening		1.30	12.308	↓	1,000	510		1,510	2,025
0400	6'-0" x 10'-0" opening, 3' high transom		1.10	14.545	Pr.	1,225	600		1,825	2,450
0420	7'-0" x 7'-0" opening		1	16	"	1,025	660		1,685	2,325
0500	Wide stile, 2'-6" x 7'-0" opening		2	8	Ea.	855	330		1,185	1,550
0520	3'-0" x 7'-0" opening		2	8		845	330		1,175	1,525
0540	3'-6" x 7'-0" opening		2	8		915	330		1,245	1,600
0560	5'-0" x 7'-0" opening		2	8	↓	1,350	330		1,680	2,075
0580	6'-0" x 7'-0" opening		1.30	12.308	Pr.	1,375	510		1,885	2,425
0600	7'-0" x 7'-0" opening	↓	1	16	"	1,475	660		2,135	2,825
1100	For full vision doors, with 1/2" glass, add				Leaf	55%				
1200	For non-standard size, add					67%				
1300	Light bronze finish, add					36%				
1400	Dark bronze finish, add					18%				
1500	For black finish, add					36%				
1600	Concealed panic device, add					1,025			1,025	1,125
1700	Electric striker release, add				Opng.	262			262	288
1800	Floor check, add				Leaf	775			775	855
1900	Concealed closer, add				"	515			515	570
2000	Flush 3' x 7' Insulated, 12" x 12" lite, clear finish	2 Sswk	2	8	Ea.	1,050	330		1,380	1,750

08 11 63 – Metal Screen and Storm Doors and Frames

08 11 63.23 Aluminum Screen and Storm Doors and Frames

		Crew	Daily Output	Labor-Hours	Unit	Material	Labor	Equipment	Total	Total Incl O&P
0010	**ALUMINUM SCREEN AND STORM DOORS AND FRAMES**									
0020	Combination storm and screen									
0400	Clear anodic coating, 6'-8" x 2'-6" wide	2 Carp	15	1.067	Ea.	162	39		201	240
0420	2'-8" wide		14	1.143		191	42		233	276
0440	3'-0" wide	↓	14	1.143	↓	191	42		233	276
0500	For 7' door height, add					5%				
1000	Mill finish, 6'-8" x 2'-6" wide	2 Carp	15	1.067	Ea.	216	39		255	298
1020	2'-8" wide		14	1.143		216	42		258	305
1040	3'-0" wide	↓	14	1.143		234	42		276	325
1100	For 7'-0" door, add					5%				
1500	White painted, 6'-8" x 2'-6" wide	2 Carp	15	1.067		262	39		301	350
1520	2'-8" wide		14	1.143		223	42		265	310
1540	3'-0" wide	↓	14	1.143		258	42		300	350
1600	For 7'-0" door, add				↓	5%				
2000	Wood door & screen, see division 08 14 33.20									

Figure 9.81

Credit: *Means Building Construction Cost Data 2007*

TELEPHONE QUOTATION

PROJECT	Office Building	DATE
		TIME
FIRM QUOTING		PHONE ( )
ADDRESS		BY
ITEM QUOTED		RECEIVED BY EBW

WORK INCLUDED	AMOUNT OF QUOTATION
Window/Curtain Wall	1 1 0 2 6 4 0 00
25,060 S.F. @ 44.00	
Tax 5% (On Material)	5 5 1 3 2 00

DELIVERY TIME 18 Weeks	TOTAL BID	1 1 5 8 0 0 0 00

DOES QUOTATION INCLUDE THE FOLLOWING: — If ☐ NO is checked, determine the following:

STATE & LOCAL SALES TAXES	☒ YES ☐ NO	MATERIAL VALUE	
DELIVERY TO THE JOB SITE	☒ YES ☐ NO	WEIGHT	
COMPLETE INSTALLATION	☒ YES ☐ NO	QUANTITY	
COMPLETE SECTION AS PER PLANS & SPECIFICATIONS	☒ YES ☐ NO	DESCRIBE BELOW	

EXCLUSIONS AND QUALIFICATIONS

ADDENDA ACKNOWLEDGEMENT	TOTAL ADJUSTMENTS
	ADJUSTED TOTAL BID

ALTERNATES

ALTERNATE NO. 1 – Custom color		8 0 0 0 0 00
ALTERNATE NO. 2 – Grid texture		1 4 0 4 0 0 00
ALTERNATE NO.		
ALTERNATE NO.		
ALTERNATE NO.		
ALTERNATE NO.		
ALTERNATE NO.		

Figure 9.82

Division 9: Finishes

Some buildings today are built "on spec," or speculatively, before they are partially or fully tenanted. In this case, interior work (primarily finishes, and electrical and mechanical distribution) is not usually completed until a tenant is secured. "Interior contractors" are becoming more prevalent in the industry. This type of firm may perform or subcontract all work in Division 9, and may or may not be a conventional general contractor that builds structures from the ground up. Because of the skills involved and the quality required, subcontractors usually specialize in only one type of finish. Hence, most finish work is subcontracted.

In today's fireproof and fire-resistant types of construction, some finish materials may be the only combustibles used in a building project. Most building codes (and specifications) require strict adherence to maximum fire, flame spread, and smoke generation characteristics. The estimator must be sure that all materials meet the specified requirements. Materials may have to be treated for fire retardancy at an additional cost.

Lathing & Plastering

The different types of plaster work require varied pricing strategies. Large open areas of continuous walls or ceilings will require considerably less labor per unit of area than small areas or intricate work such as archways, curved walls, cornices, and at window returns. Gypsum and metal lath are most often used as subbases. However, plaster may be applied directly on masonry, concrete, and in some restoration work, wood. In the latter cases, a bonding agent may be specified.

The number of coats of plaster may also vary. Traditionally, a scratch coat is applied to the substrate. A brown coat is then applied two days later, and the finish, smooth coat seven days after the brown coat. Currently, the systems most often used are two-coat and one-coat (imperial plaster on "blueboard"). Textured surfaces, with and without patterns, may be required. All of these variables in plaster work make it difficult to develop "system" prices. Each project, and even areas within each project, must be examined individually.

The quantity takeoff should proceed in the normal construction sequence—furring (or studs), lath, plaster, and accessories. Studs, furring, and/or ceiling suspension systems, whether wood or steel, should be taken off separately. Responsibility for the installation of these items should be made clear. Depending on local work practices, lathers may or may not install studs or furring. These materials are usually estimated by the piece or linear foot, and sometimes by the square foot. Lath is traditionally estimated by the square yard (for both gypsum and metal lath), but is more recently taken off by the square foot. Usually, a 5% allowance for waste is included. Casing bead, corner bead, and other accessories are measured by the linear foot. An extra foot of surface area should be allowed for each linear foot of corner or stop. Although wood plaster grounds are usually installed by carpenters, they should be measured when taking off the plaster requirements. Plastering is also traditionally measured by the square yard. Deductions for openings vary by preference—from zero deduction to 50% of all openings over two feet in width. Some estimators deduct a percentage of the total yardage for openings. The estimator should allow one extra square foot of wall area for each linear foot of inside or outside corner located below the ceiling level. Also, double the areas of small radius work. Quantities are determined by measuring surface areas (walls, ceilings). The estimator must consider both the complexity and the intricacy of the work, and in pricing plaster work, should also consider quality. Basically, there are two quality categories:

1. Ordinary—for commercial purposes, and with waves 1/8" to 3/16" in 10 feet, angles and corners fairly true.
2. First Quality—with variations less than 1/16" in 10 feet. Labor costs for first quality work are approximately 20% more than for ordinary plastering.

Drywall

With the advent of light-gauge metal framing, tin snips are as important to the carpenter as the circular saw. Metal studs and framing are usually installed and included by the drywall subcontractor. The estimator should make sure that studs (and other framing—whether metal or wood) are not included twice by different subcontractors. In some drywall systems, such as shaftwall, the framing is integral and installed simultaneously with the drywall panels.

Metal studs are manufactured in various widths (1-5/8", 2-1/2", 3-5/8", 4", and 6") and in various gauges, or metal thicknesses. They may be used for both load-bearing and non-load-bearing partitions, depending on design criteria and code requirements. Metal framing is particularly useful due to the prohibitive use of structural wood (combustible) materials in new building construction. Metal studs, track, and accessories are purchased by the linear foot, and usually stocked in 8' to 16' lengths, by 2' increments. For large orders, metal studs can be purchased in any length up to 20'.

For estimating, light-gauge metal framing is taken off by the linear foot or by the square foot of wall area of each type. Different wall types—with different stud widths, stud spacing, or drywall requirements—should each be taken off separately, especially if estimating by the square foot.

Metal studs can be installed quickly. Depending on the specification, metal studs may have to be fastened to the track with self-tapping screws, tack welds, or clips, or may not have to be prefastened. Each condition will affect the labor costs. Fasteners, such as screws, clips, and powder-actuated studs, are expensive, though labor-saving. These costs must be included.

Drywall may be purchased in various thicknesses—1/4" to 1"—and in various sizes— 2' x 8' to 4' x 20'. Different types include standard, fire-resistant, water-resistant, blueboard, coreboard, and pre-finished. There are many variables and possible combinations of sizes and types. While the installation cost of 5/8" standard drywall may be the same as that of 5/8"

Sample Estimate: Division 9

Most of the finishes for the sample project involve large quantities in wide open areas. For an interior space broken up into many rooms or suites, the estimate would be much more involved, and a detailed Room Finish Schedule should be provided. The estimate sheets for the sample project are shown in Figures 9.85 to 9.88. On all the estimate sheets for the project, items are entered on every other line. Many more sheets are used this way, but paper is cheap. Invariably, omitted items must be inserted at the last minute, and sheet subtotals and division totals will be recalculated as a result. If the sheets are concise, neat, and organized, recalculation can be easy and will involve less chance of error.

Particularly when estimating drywall, the same dimensions can be used to calculate the quantities of different items—in this case, drywall, metal studs, and accessories. When an opening (to be finished with drywall) is deducted from the drywall surface area, perimeter dimensions are used to calculate corner bead. Similarly, at joints with items such as windows, the same dimensions are used for J-bead. The quantity sheet for drywall and associated items is shown in Figure 9.89. Types, sizes, and different applications of drywall and studs are listed (and priced) separately. The square foot quantities of studs and drywall can be used for comparison to check for possible errors. When such a method is used, the estimator must be aware of those partitions which receive drywall only on one side, as opposed to both sides, and those with more than one layer. A comparison for the sample project is performed as follows:

Framing	Stud Area	No. of Sides	No. of Layers	Drywall Area
6"	300 SF	2	1	600 SF
3-5/8"	3,560 SF	2	1	7,120 SF
3-5/8"	10,104 SF	1	1	10,104 SF
1-5/8"	5,640 SF	1	2	11,280 SF
1-5/8"	1,704 SF	1	1	1,704 SF
Furring	4,200 SF	1	1	4,200 SF
	25,508 SF			35,008 SF

$$25,508 \text{ SF} = 35,008 - (300 + 3,560 + 5,640) = 25,508 \text{ SF}$$

Note that furring is included in the above cross-check calculations as square feet. The estimator must always be aware of units during *any* calculations. When deductions are made from the drywall totals for two-side application and double layers, the drywall and stud quantities should equate.

For the ceiling grid, because of the large, open areas, no allowance for waste is included. Note, however, that the dimensions used are for the exterior of the building (210' × 90' × 3 floors). This includes a small overage. For purchasing purposes, all ceiling perimeter dimensions would be measured separately for edge moldings. A 5% allowance is added to the ceiling tile. The large quantity of light fixtures must be considered and is deducted from the total. Deductions for light fixtures should be based on experience. Carpeting quantities are also derived from the building perimeter measurements, with appropriate deductions for the building core.

Sample Estimate: Division 9

CONSOLIDATED ESTIMATE

PROJECT: Office Building
LOCATION:
TAKE OFF BY: ABC

CLASSIFICATION: Division 9
ARCHITECT:
QUANTITIES BY: ABC
PRICES BY: ABC

SHEET NO. 1 of 4
ESTIMATE NO:
DATE: Jan-07
CHECKED BY: GHI

DESCRIPTION	SOURCE	QUANT	UNIT	MATERIAL COST	MATERIAL TOTAL	LABOR COST	LABOR TOTAL	EQUIPMENT COST	EQUIPMENT TOTAL	SUBCONTRACT COST	SUBCONTRACT TOTAL	TOTAL COST	TOTAL TOTAL
Division 9: Finishes													
Framing : Metal Studs													
6" wide - 25 gauge, 16" O.C.	09 22 16.13 2500	300	S.F.	0.58	174	0.62	186						
3-5/8" wide - 25 gauge, 16" O.C.	09 22 16.13 2300	13664	S.F.	0.39	5329	0.61	8335						
1-5/8" wide - 25 gauge, 16" O.C.	09 22 16.13 2000	7344	S.F.	0.28	2056	0.59	4333						
Accessories													
7/8" Furring Channel	09 29 15.10 0900	31.5	C.L.F.	26.50	835	113.00	3560						
J Trim, galvanized steel, 5/8" wide	09 29 15.10 1120	34.1	C.L.F.	22.00	750	99.50	3393						
Corner Bead, galvanized steel	09 29 15.10 0300	58.4	C.L.F.	13.50	788	73.50	4292						
Drywall													
5/8" F.R. @Columns	09 29 10.30 4050	5640	S.F.	0.84	4738	1.96	11054						
5/8" F.R. @Core	09 29 10.30 2150	7720	S.F.	0.42	3242	0.61	4709						
5/8" Standard	09 29 10.30 2050	16008	S.F.	0.43	6883	0.61	9765						
Shaftwall: @ Elevator													
5/8" Fire Rated Gypsum Board	09 21 16.23 0300	2040	S.F.	1.62	3305	3.26	6650						
Subtotals				$	28,101	$	56,278						

To download this and other forms in this book, visit www.rsmeans.com/supplement/67303B.asp

CONSOLIDATED ESTIMATE

SHEET NO. 2 of 4

PROJECT: Office Building
LOCATION:
TAKE OFF BY: ABC   QUANTITIES BY: ABC   PRICES BY: DEF   EXTENSIONS BY: DEF
CLASSIFICATION: Division 9
ARCHITECT: As Shown

ESTIMATE NO:
DATE: Jan-07
CHECKED BY: GHI

DESCRIPTION	SOURCE	QUANT	UNIT	MATERIAL COST	MATERIAL TOTAL	LABOR COST	LABOR TOTAL	EQUIPMENT COST	EQUIPMENT TOTAL	SUBCONTRACT COST	SUBCONTRACT TOTAL	TOTAL COST	TOTAL
Division 9: (Cont'd)													
Fireproofing: @ Beams	07 81 16.10 0400	40500	S.F.							1.35	54675		
Total Fireproofing										$	54,675		
Ceramic Tile:													
Walls, Thin Set	09.30 13.10 5400	1584	S.F.							6.40	10138		
Bull Nose, Thin Set	09.30 13.10 2500	396	L.F.							9.05	3584		
Cove Base, Thin Set	09.30 13.10 0700	396	L.F.							9.80	3881		
Floors, Porcelain Type	09.30 13.10 3300	960	S.F.							9.35	8976		
Total Ceramic Tile										$	26,578		
Accoustical Ceiling:													
Grid	09 53 23.30 0050	56700	S.F.	0.55	31185	0.37	20979						
Tiles, Mineral Fiber, 24" x 24"	09 51 23.10 3740	53735	S.F.	1.91	102634	0.51	27405						
Total Ceiling				$	133,819	$	48,384						

Figure 9.86

264

To download this and other forms in this book, visit www.rsmeans.com/supplement/67303B.asp

CONSOLIDATED ESTIMATE

PROJECT: Office Building   CLASSIFICATION: Division 9   SHEET NO. 3 of 4

LOCATION: ARCHITECT: ESTIMATE NO:

TAKE OFF BY: ABC   QUANTITIES BY: ABC   PRICES BY: ABC   EXTENSIONS BY: DATE: Jan-07   CHECKED BY: GHI

DESCRIPTION	SOURCE		QUANT	UNIT	MATERIAL		LABOR		EQUIPMENT		SUBCONTRACT		TOTAL	
					COST	TOTAL	COST	TOTAL	COST	TOTAL	COST	TOTAL	COST	TOTAL
Division 9: (Cont'd)														
Flooring														
Carpet	09 68 16.10	3200	5850	SY							48.00	280800		
Cove Base	09 65 13.13	1150	2530	L.F.							2.13	5389		
Parquet @ Lobby & Elevators	09 64 29.10	6500	1800	S.F.							9.40	16920		
Total Flooring											$	303,109		
Painting														
Walls (Drywall)	09 91 23.72	0840	27784	S.F.							0.62	17226		
Wood Doors	09 91 23.33	1800	21	Ea.							47.70	1002		
Metal Doors (Primed)	09 91 23.33	1000	14	Ea.							42.15	590		
Block Walls	09 91 23.72	2880	9846	S.F.							0.47	4628		
Total Painting											$	23,446		

Figure 9.87

Sample Estimate: Division 9

CONSOLIDATED ESTIMATE

PROJECT: Office Building	SHEET NO. 4 of 4
LOCATION:	ESTIMATE NO:
TAKE OFF BY: ABC	DATE: Jan-07
CLASSIFICATION: Division 9	CHECKED BY: GHI
ARCHITECT:	
QUANTITIES BY: ABC	PRICES BY:	EXTENSIONS BY: DEF

SOURCE	DESCRIPTION	QUANT	UNIT	MATERIAL DEF COST	MATERIAL TOTAL	LABOR COST	LABOR TOTAL	EQUIPMENT DEF COST	EQUIPMENT TOTAL	SUBCONTRACT COST	SUBCONTRACT TOTAL	TOTAL COST	TOTAL TOTAL
	Division 9: (Cont'd)												
	Sheet 1: Drywall & Framing				$ 28,101		$ 56,278						
	Sheet 2: Fireproofing										$ 54,675		
	Ceramic Tile										$ 26,578		
	Accoustical Ceiling				$ 133,819		$ 48,384						
	Sheet 3: Flooring										$ 303,109		
	Painting										$ 23,446		
	Division 9 Totals				$ 161,920		$ 104,662				$ 407,808		

Figure 9.88

To download this and other forms in this book, visit www.rsmeans.com/supplement/67303B.asp

QUANTITY SHEET

PROJECT: Office Building   Division 9   SHEET NO. 1 of 1
LOCATION: ARCHITECT: ESTIMATE NO:
TAKE OFF BY: EXTENSIONS BY: DATE:
    CHECKED BY:

DESCRIPTION	NO.	DIMENSIONS L	DIMENSIONS W	DIMENSIONS H	Studs	UNIT	5/8" Std. Drywall	UNIT	5/8" F.R. Drywall	UNIT	Acc.	UNIT
Partitions - 25 ga.												
Bath Chase 6"	3	10'		10'	300	SF			600	SF		
Interior 3 5/8"												
Lobby	1	86'		10'	860	SF			1720	SF		
Core	3	90'		10'	2700	SF			5400	SF		
Corner Bead	27										270	LF
Furring 16" O.C.	3	140'		10'			4200	SF				
Exterior 3 5/8"	3	560'		10'	10104	SF	10104	SF			3151	LF
Deduct Windows	(	496'		4'-5"	{ 10104 }	SF	{ }	SF				
Corner Bead	6	496'									3408	LF
J - Bead	96	4.5'										
	6	496'										
	96	4.5'										
Window Returns		3408'	0.5'		1704	SF	1704	SF			3408	LF
1 5/8" Studs												
Columns 1 5/8"	3	188'		10'	5640	SF			5640	SF		
Unfinished									5640	SF		
Taped									5640	SF		
Corner Bead	3	72'		10'							2160	LF
Quantity Summary												
Studs: 6"					300	SF						
3 5/8"					13664	SF						
1 5/8"					7344	SF						
Drywall: 5/8" F.R.									13360	SF		
Unfinished: 5/8" F.R.									5640	SF		
5/8" Std.							16008	SF				
Corner Bead											5838	LF
J - Bead											3408	LF
Furring											3151	LF

Figure 9.89

fixtures. If detail is required, the estimator should measure and record the lengths of each size of pipe. When a fitting is crossed, a list of all fitting types and sizes can be used for counting. Colored pencils can be used to mark runs that have been completed, and to note termination points on the main line where measurements are stopped so that a branch may be taken off. Care must be taken to note changes in pipe material. Since different materials can only meet at a joint, it should become an automatic habit to see that the piping material going into a joint is the same as that leaving the joint.

When summarizing quantities of piping for estimate or purchase, it is good practice to round the totals of each size up to the lengths normally available from the supply house or mill. This method is even more appropriate in larger projects, where rounding might be done to the nearest hundred or thousand feet. With this approach, a built-in percentage for scrap or waste is allowed. Rigid copper tubing and rigid plastic pipe are normally supplied in 20' lengths, cast iron soil pipe in either 5' or 10' sections. Steel pipe of 2" diameter or less is furnished in 21' lengths, and pipe of a larger diameter is available in single random lengths ranging from 16' to 22'.

Apart from counting and pricing every pipe and fitting, there are other ways to determine budget costs for plumbing. *Means Building Construction Cost Data* provides costs for fixtures, as well as the associated costs for rough-in of the supply, waste, and vent piping as shown in Figure 9.95. The rough-in costs include piping within 10' of the fixture. If these prices are used, further costs must be added for the stacks and mains. Another method involves adding percentages to the cost of the fixtures. Recommended percentages are shown in Figure 9.96. Using information such as that provided in Figures 9.95 and 9.96, the estimator can develop systems to be used for budget pricing.

In addition to the cost of pipe installation, the estimator must also consider any associated costs. In many cases, mechanical and electrical subcontractors must dig, by hand, their own under-slab trenches. Site utility excavation (and backfill) may also be included, if required. Underground piping may require special wrapping. For interior piping, the estimator must visualize the installation in order to determine how—and to what—the pipe hangers are attached. Overhead installations will require rolling scaffolding. For installations higher than an average of 15', labor costs may be increased by the following suggested percentages:

Ceiling Height	Labor Increase
15'–20'	10%
20'–25'	20%
25'–30'	30%
30'–35'	40%
35'–40'	50%
Over 40'	60%

Fire Protection

The takeoff of fire protection systems (sprinklers and standpipes) is very much like that of other plumbing—the estimator should measure the pipe loops and count fittings, valves, sprinkler heads, alarms, and other components. The estimator then makes note of special requirements, as well as any conditions that would affect job performance and cost.

There are many different types of sprinkler systems; examples are wet pipe, dry pipe, pre-action, and chemical. Each of these types involves a different set of requirements and costs. Figure 9.97, from *Means Mechanical Cost Data*, provides descriptions of various sprinkler systems.

Most sprinkler systems must be approved by Factory Mutual for insurance purposes *before* installation. This requirement means that shop drawings must be produced and submitted for approval. The shop drawings take time, and the approval process can be drawn out even more. This process can cause serious delays and added expense if it is not anticipated and given adequate consideration at the estimating and scheduling phase.

Square foot historical costs for fire protection systems may be developed for budget purposes. These costs may be based on the relative hazard of occupancy—light, ordinary, and extra. A comparison of some requirements of the different hazards is shown in Figure 9.98. Consideration must also be given to special or unusual requirements. For example, many architects specify that sprinkler heads must be located in the center of ceiling tiles. Each head may require extra elbows and nipples for precise location. Recessed heads are more expensive. Special dry pendant heads are required in areas subject to freezing. When installing a sprinkler system in an existing structure, a completely new water service may be required in addition to the existing domestic water service. These are just a few examples of requirements that may necessitate an adjustment of square foot costs.

Heating, Ventilating, & Air Conditioning

As with plumbing, equipment for HVAC should be taken off first so that the estimator is familiarized with the various systems and layouts. While actual pieces of equipment must be counted at this time, it is also necessary to note sizes, capacities, controls, and special characteristics and features. The weight and size of equipment may be important if the unit is especially large or is going into comparatively close quarters. If hoisting or rigging is not included in the subcontract price, then costs for placing the equipment must be figured and listed. From the equipment totals, other important items, such as motor starters, valves, strainers, gauges, thermometers, traps, and air vents, can also be counted.

Sheet metal ductwork for heating, ventilating, and air conditioning is usually estimated by weight. The lengths of the various sizes are measured and recorded on a worksheet. The weight per foot of length is then determined. Figure 9.99 is a conversion chart for determining the weight of ductwork based on size and material. A count must also be made of all duct-associated accessories, such as fire dampers, diffusers, and registers. This count may be done during the duct takeoff. It is usually less confusing, however, to make a separate count. For budget purposes, ductwork, insulation, diffusers, and registers may be estimated using the information in Figure 9.100.

The takeoff of heating, ventilating, and air conditioning pipe and fittings is accomplished in a manner similar to that of plumbing. In addition to the general, miscellaneous items noted during the review of plans and specifications, the heating, ventilating, and air conditioning estimate usually includes the work of subcontractors. All material suppliers and appropriate subcontractors should be notified as soon as possible to verify material availability and pricing and to ensure timely submission of bids. Typical subcontract work includes:

- Balancing
- Controls
- Insulation
- Water treatment
- Sheet metal
- Core drilling

If similar systems are used repeatedly, it is easy to develop historical square foot costs for budget and comparison purposes. However, with variation in the types and size of HVAC systems, the estimator must use caution when using square foot prices. A way to adjust such relative prices can be to adjust costs based on the quality or complexity of the installation:

Quality/Complexity	Adjustment
Economy/Low	0 to 5%
Good/Medium	5 to 15%
Above Average/High	15 to 25%

Plumbing Fixture Installation Time

Item	Rough-In	Set	Total Hrs.	Item	Rough-In	Set	Total Hrs.
Bathtub	5	5	10	Shower head only	2	1	3
Bathtub and shower, cast iron	6	6	12	Shower drain	3	1	4
Fire hose reel and cabinet	4	2	6	Shower stall, slate		15	15
Floor drain to 4" diameter	3	1	4	Slop sink	5	3	8
Grease trap, single, cast iron	5	3	8	Test 6 fixtures			14
Kitchen gas range		4	4	Urinal, wall	6	2	8
Kitchen sink, single	4	4	8	Urinal, pedestal or floor	6	4	10
Kitchen sink, double	6	6	12	Water closet and tank	4	3	7
Laundry tubs	4	2	6	Water closet and tank, wall hung	5	3	8
Lavatory, wall-hung	5	3	8	Water heater, 45 gals. gas, automatic	5	2	7
Lavatory, pedestal	5	3	8	Water heaters, 65 gals. gas, automatic	5	2	7
Shower and stall	6	4	10	Water heaters, electric, plumbing only	4	2	6

Figure 9.94

22 41 Residential Plumbing Fixtures

22 41 39 – Residential Faucets, Supplies, and Trim

22 41 39.10 Faucets and Fittings		Crew	Daily Output	Labor-Hours	Unit	Material	2007 Bare Costs Labor	2007 Bare Costs Equipment	Total	Total Incl O&P
2000	Laundry faucets, shelf type, IPS or copper unions	1 Plum	12	.667	Ea.	45	30		75	94.50
2100	Lavatory faucet, centerset, without drain	↓	10	.800	↓	40	36		76	98
2210	Porcelain cross handles and pop-up drain									
2220	Polished chrome	1 Plum	6.66	1.201	Ea.	138	54		192	233
2230	Polished brass	"	6.66	1.201	"	208	54		262	310
2260	Single lever handle and pop-up drain									
2270	Black nickel	1 Plum	6.66	1.201	Ea.	220	54		274	325
2280	Polished brass		6.66	1.201		220	54		274	325
2290	Polished chrome		6.66	1.201		161	54		215	258
2810	Automatic sensor and operator, with faucet head		6.15	1.301		330	58.50		388.50	455
3000	Service sink faucet, cast spout, pail hook, hose end		14	.571		76.50	25.50		102	123
4000	Shower by-pass valve with union		18	.444		54	19.90		73.90	89.50
4200	Shower thermostatic mixing valve, concealed	↓	8	1	↓	246	45		291	340
4220	Shower pressure balancing mixing valve,									
4230	With shower head, arm, flange and diverter tub spout									
4240	Chrome	1 Plum	6.14	1.303	Ea.	141	58.50		199.50	243
4250	Polished brass		6.14	1.303		198	58.50		256.50	305
4260	Satin		6.14	1.303		198	58.50		256.50	305
4270	Polished chrome/brass		6.14	1.303		162	58.50		220.50	266
5000	Sillcock, compact, brass, IPS or copper to hose	↓	24	.333	↓	6.65	14.95		21.60	30

22 42 Commercial Plumbing Fixtures

22 42 13 – Commercial Water Closets, Urinals, and Bidets

22 42 13.30 Urinals

	URINALS	Crew	Daily Output	Labor-Hours	Unit	Material	Labor	Equipment	Total	Total Incl O&P
0010	**URINALS**									
3000	Wall hung, vitreous china, with hanger & self-closing valve									
3100	Siphon jet type	Q-1	3	5.333	Ea.	325	215		540	685
3120	Blowout type		3	5.333		360	215		575	720
3300	Rough-in, supply, waste & vent		2.83	5.654		176	228		404	540
5000	Stall type, vitreous china, includes valve		2.50	6.400		540	258		798	985
6980	Rough-in, supply, waste and vent		1.99	8.040	↓	244	325		569	760

22 42 13.40 Water Closets

	WATER CLOSETS	Crew	Daily Output	Labor-Hours	Unit	Material	Labor	Equipment	Total	Total Incl O&P
0010	**WATER CLOSETS**									
3000	Bowl only, with flush valve, seat									
3100	Wall hung	Q-1	5.80	2.759	Ea.	315	111		426	510
3200	For rough-in, supply, waste and vent, single WC		2.56	6.250		520	252		772	950
3300	Floor mounted		5.80	2.759		375	111		486	580
3350	With wall outlet		5.80	2.759		535	111		646	755
3400	For rough-in, supply, waste and vent, single WC		2.84	5.634	↓	248	227		475	615

22 42 16 – Commercial Lavatories and Sinks

22 42 16.10 Handwasher-Dryer Module

	HANDWASHER-DRYER MODULE	Crew	Daily Output	Labor-Hours	Unit	Material	Labor	Equipment	Total	Total Incl O&P
0010	**HANDWASHER-DRYER MODULE**									
0030	Wall mounted									
0040	With electric dryer									
0050	Sensor operated	Q-1	8	2	Ea.	2,300	80.50		2,380.50	2,650
0110	Sensor operated (ADA)	↓	8	2		2,175	80.50		2,255.50	2,525
0140	Sensor operated (ADA), surface mounted	↓	8	2	↓	2,825	80.50		2,905.50	3,225
0150	With paper towels									
0180	Sensor operated	Q-1	8	2	Ea.	2,250	80.50		2,330.50	2,600

Figure 9.95

Plumbing Approximations for Quick Estimating

Water Control:
Water Meter; Backflow Preventer,
Shock Absorbers; Vacuum Breakers; } . . . . 10 to 15% of Fixtures
Mixer.

Pipe and Fittings: . . . . 30 to 60% of Fixtures

> **Note:** Lower percentage for compact buildings or larger buildings with plumbing in one area.
> Larger percentage for large buildings with plumbing spread out.
> In extreme cases pipe may be more than 100% of fixtures.
> Percentages **do not** include special purpose or process piping.

Plumbing Labor:
1 & 2 Story Residential . . . . .Rough-in Labor = 80% of Materials
Apartment Buildings . . . . Rough-in Labor = 90 to 100% of Materials
Labor for handling and placing fixtures is approximately 25 to 30% of fixtures.

Quality/Complexity Multiplier (For all installations)
Economy installation, add . . . . .0 to 5%
Good quality, medium complexity, add. . . . . .5 to 15%
Above average quality and complexity, add . . . . .15 to 25%

Figure 9.96

R211313-10   Sprinkler Systems (Automatic)

Sprinkler systems may be classified by type as follows:

1. **Wet Pipe System.** A system employing automatic sprinklers attached to a piping system containing water and connected to a water supply so that water discharges immediately from sprinklers opened by a fire.

2. **Dry Pipe System.** A system employing automatic sprinklers attached to a piping system containing air under pressure, the release of which as from the opening of sprinklers permits the water pressure to open a valve known as a "dry pipe valve". The water then flows into the piping system and out the opened sprinklers.

3. **Pre-Action System.** A system employing automatic sprinklers attached to a piping system containing air that may or may not be under pressure, with a supplemental heat responsive system of generally more sensitive characteristics than the automatic sprinklers themselves, installed in the same areas as the sprinklers; actuation of the heat responsive system, as from a fire, opens a valve which permits water to flow into the sprinkler piping system and to be discharged from any sprinklers which may be open.

4. **Deluge System.** A system employing open sprinklers attached to a piping system connected to a water supply through a valve which is opened by the operation of a heat responsive system installed in the same areas as the sprinklers. When this valve opens, water flows into the piping system and discharges from all sprinklers attached thereto.

5. **Combined Dry Pipe and Pre-Action Sprinkler System.** A system employing automatic sprinklers attached to a piping system containing air under pressure with a supplemental heat responsive system of generally more sensitive characteristics than the automatic sprinklers themselves, installed in the same areas as the sprinklers; operation of the heat responsive system, as from a fire, actuates tripping devices which open dry pipe valves simultaneously and without loss of air pressure in the system. Operation of the heat responsive system also opens

approved air exhaust valves at the end of the feed main which facilitates the filling of the system with water which usually precedes the opening of sprinklers. The heat responsive system also serves as an automatic fire alarm system.

6. **Limited Water Supply System.** A system employing automatic sprinklers and conforming to these standards but supplied by a pressure tank of limited capacity.

7. **Chemical Systems.** Systems using halon, carbon dioxide, dry chemical or high expansion foam as selected for special requirements. Agent may extinguish flames by chemically inhibiting flame propagation, suffocate flames by excluding oxygen, interrupting chemical action of oxygen uniting with fuel or sealing and cooling the combustion center.

8. **Firecycle System.** Firecycle is a fixed fire protection sprinkler system utilizing water as its extinguishing agent. It is a time delayed, recycling, preaction type which automatically shuts the water off when heat is reduced below the detector operating temperature and turns the water back on when that temperature is exceeded. The system senses a fire condition through a closed circuit electrical detector system which controls water flow to the fire automatically. Batteries supply up to 90 hour emergency power supply for system operation. The piping system is dry (until water is required) and is monitored with pressurized air. Should any leak in the system piping occur, an alarm will sound, but water will not enter the system until heat is sensed by a firecycle detector.

Area coverage sprinkler systems may be laid out and fed from the supply in any one of several patterns as shown below. It is desirable, if possible, to utilize a central feed and achieve a shorter flow path from the riser to the furthest sprinkler. This permits use of the smallest sizes of pipe possible with resulting savings.

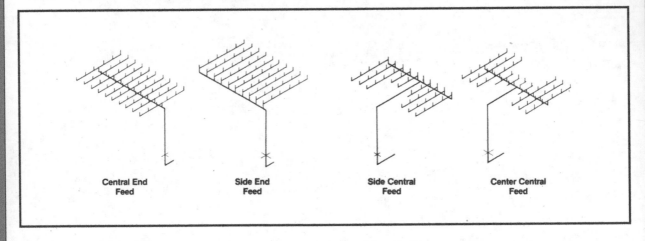

| Central End Feed | Side End Feed | Side Central Feed | Center Central Feed |

Credit: Means Mechanical Cost Data 2007

Figure 9.97

Fixtures should be taken off using the fixture schedule in conjunction with the reflected ceiling plan. Fixture counts should have outlet boxes, plaster rings, or Greenfield with connectors and fixture wire if needed. Include fixture supports as needed.

Separate quantity sheets should be used for each of the major categories of electrical work. By keeping each system on different sheets, the estimator will find it easier to isolate various costs. This format becomes a reference for purchasing and cost control. It also provides a breakdown of items that can be submitted to the general contractor when billing.

While making takeoffs of each category, identify all other required components of the systems. Determine the materials that will be part of a quotation or sub-bid from a supplier, including additional items to complete the vendor's package.

Switchgear

Material costs for large equipment, such as switchgear motor control centers and associated items, should be obtained from suppliers or manufacturers. When determining the installation cost for large equipment (for any division), the estimator must consider a number of factors to be sure that all of the following work is included:

- Access and placement
- Uncrating
- Rigging and setting
- Pads and anchors
- Leveling and shimming
- Assembly of components
- Connections
- Temporary protection requirements

While pads and anchors for large equipment may be well-designated on the plans, supports for smaller equipment, such as panels and transformers, may not be well-defined. For example, floor-to-ceiling steel supports may be required. The same consideration must be given to large cable troughs and conduits. Special support requirements may be necessary. If so, they must be included, whether they are the responsibility of the electrical or general contractor. For elevated installation of equipment, labor costs should be adjusted. Suggested adjustments follow:

Ceiling Height	Labor Adjustment
10'–15'	+ 15%
16'–25'	+ 30%
Over 25'	+ 35%

Ducts & Cable Trays

Bus ducts and cable trays should be estimated by component:

- Type and size of duct or tray
- Material (aluminum or galvanized)
- Hangers
- Fittings

The installation of under-floor duct systems (and conduit) must be scheduled after placement of any reinforcing at the bottom of slabs, but before installation of the upper steel. Without proper coordination, delays can occur. The estimate and preliminary schedule should reflect associated costs. Because of the high cost of tray and duct systems, the takeoff should be as accurate as possible. Fittings should not be deducted from straight lengths. In this way, an adequate allowance for waste should be provided. For high ceiling installations, labor costs for bus duct and cable trays should be adjusted accordingly. Suggested adjustments are:

Ceiling Height	Labor Adjustment
16'–20'	+10%
21'–25'	+20%
26'–30'	+25%
31'–35'	+30%
36'–40'	+35%
Over 40'	+40%

Feeders

Feeder conduit and wire should be carefully taken off using a scale or printed dimensions. This is a more accurate method than the use of a rotometer. Large conduit and wire are expensive and require a considerable amount of labor. Accurate quantities are important. Switchboard locations should be marked on each floor before measuring horizontal runs. The distance between floors should be marked on riser plans and added to the horizontal for a complete feeder run. Conduit should be measured at right angles to the structure unless it has been determined that it can be routed directly. Elbows, terminations, bends, or expansion joints should also be taken off at this time. Under-slab and high ceiling installations should be treated the same as in the case of ducts and trays. The estimator should also consider the weight of material for high ceiling installations. Extra workers may be required for heavier components. Typical weights are shown in Figure 9.109. If standard feeder takeoff sheets are used, the wire column should reflect longer lengths than the conduit. This is because of added amounts of wire used in the panels and switchboard to make connections.

If the added length is not shown on the plans, it can be determined by checking a manufacturer's catalog. Conduit, cable supports, and accessories should all be totalled at this time.

Branch Circuits

Branch circuits may be taken off using a rotometer. The estimator should take care to start and stop accurately at boxes. Start with two wire circuits and mark with colored pencil as items are taken off. Add about 5% to conduit quantities for normal waste. On wire, add 10% to 12% overage to make connections. Add conduit fittings, such as locknuts, bushings, couplings, expansion joints, and fasteners. Two conduit terminations per box are average. Figure 9.110 shows prices for conduit as presented in *Means Building Construction Cost Data*. Note that a minimum amount of fittings are included in the costs.

Wiring devices should be entered with plates, boxes, and plaster rings. Calculate stub ups or drops for wiring devices. Switches should be counted and multiplied by the distance from switch to ceiling.

Receptacles are handled similarly, depending on whether they are wired from the ceiling or floor. In many cases, there are two conduits going to a receptacle box as you feed in and out to the next outlet. Some estimators let rotometers overrun outlets purposely as an adjustment for vertical runs; this practice can, however, lead to inaccuracies.

In metal stud partitions, it is often acceptable to go horizontally from receptacle to receptacle. In wood partitions, horizontal runs are also the usual practice. In suspended ceilings, the specifications should be checked to see if straight runs are allowed. If the space above the ceiling is used as a return air plenum, conduit is usually required for all wiring.

Motors, safety switches, starters, and controls should each include power lines, junction boxes, supports, wiring troughs, and wire terminations. Short runs of Greenfield or Sealtite with appropriate connectors and wire should be added for motor wiring.

For large installations, the economy of scale may have a definite impact on the electrical costs. If large quantities of a particular item are installed in the same general area, certain deductions can be made for labor. Suggested deductions include:

	Quantity	Labor Deduction
Under-floor ducts, bus ducts, conduit, cable systems:	150 to 250 LF	–10%
	251 to 350 LF	–15%
	351 to 500 LF	–20%
	Over 500 LF	–25%
Outlet boxes:	25 to 50	–15%
	51 to 75	–20%
	76 to 100	–25%
	Over 100	–30%
Wiring devices:	10 to 25	–20%
	26 to 50	–25%
	51 to 100	–30%
	Over 100	–35%
Lighting fixtures:	25 to 50	–15%
	51 to 75	–20%
	76 to 100	–25%
	Over 100	–30%

The estimator for the general contractor may require only a budget cost for the purpose of verifying subcontractor bids. As described in the section on mechanical estimating, square foot or systems costs developed from similar, past projects can also be useful for electrical estimating.

Typical Material Weights

Conduit Weight Comparisons (Lbs. per 100 ft.)

Type	1/2"	3/4"	1"	1-1/4"	1-1/2"	2"	2-1/2"	3"	3-1/2"	4"	5"	6"
Rigid Aluminum	28	37	55	72	89	119	188	246	296	350	479	630
Rigid Steel	79	105	153	201	249	332	527	683	831	972	1314	1745
Intermediate Steel (IMC)	60	82	116	150	182	242	401	493	573	638		
Electrical Metallic Tubing (EMT)	29	45	65	96	111	141	215	260	365	390		
Polyvinyl Chloride, Schedule 40	16	22	32	43	52	69	109	142	170	202	271	350
Polyvinyl Chloride Encased Burial						38		67	88	105	149	202
Fibre Duct Encased Burial						127		164	180	206	400	511
Fibre Duct Direct Burial						150		251	300	354		
Transite Encased Burial						160		240	290	330	450	550
Transite Direct Burial						220		310		400	540	640

Weight Comparisons of Common Size Cast Boxes in Lbs.

Size NEMA 4 or 9	Cast Iron	Cast Aluminum	Size NEMA 7	Cast Iron	Cast Aluminum
6" x 6" x 6"	17	7	6" x 6" x 6"	40	15
8" x 6" x 6"	21	8	8" x 6" x 6"	50	19
10" x 6" x 6"	23	9	10" x 6" x 6"	55	21
12" x 12" x 6"	52	20	12" x 6" x 6"	100	37
16" x 16" x 6"	97	36	16" x 16" x 6"	140	52
20" x 20" x 6"	133	50	20" x 20" x 6"	180	67
24" x 18" x 8"	149	56	24" x 18" x 8"	250	93
24" x 24" x 10"	238	88	24" x 24" x 10"	358	133
30" x 24" x 12"	324	120	30" x 24" x 10"	475	176
36" x 36" x 12"	500	185	30" x 24" x 12"	510	189

Size Required and Weight (Lbs./1000 L.F.) of Aluminum and Copper THW Wire by Ampere Load

Amperes	Copper Size	Aluminum Size	Copper Weight	Aluminum Weight
15	14	12	24	11
20	12	10	33	17
30	10	8	48	39
45	8	6	77	52
65	6	4	112	72
85	4	2	167	101
100	3	1	205	136
115	2	1/0	252	162
130	1	2/0	324	194
150	1/0	3/0	397	233
175	2/0	4/0	491	282
200	3/0	250	608	347
230	4/0	300	753	403
255	250	400	899	512
285	300	500	1068	620
310	350	500	1233	620
335	400	600	1396	772
380	500	750	1732	951

Weight (Lbs./L.F.) of 4 Pole Aluminum and Copper Bus Duct by Ampere Load

Amperes	Aluminum Feeder	Copper Feeder	Aluminum Plug-In	Copper Plug-In
225			7	7
400			8	13
600	10	10	11	14
800	10	19	13	18
1000	11	19	16	22
1350	14	24	20	30
1600	17	26	25	39
2000	19	30	29	46
2500	27	43	36	56
3000	30	48	42	73
4000	39	67		
5000		78		

Figure 9.109

26 05 26 – Grounding and Bonding for Electrical Systems

26 05 26.80 Grounding	Crew	Daily Output	Labor-Hours	Unit	Material	2007 Bare Costs Labor	Equipment	Total	Total Incl O&P
0390	Bare copper wire, stranded, #8	1 Elec	11	.727	C.L.F.	40	32		72	91.50
0400	#6		10	.800		73	35		108	133
0600	#2	2 Elec	10	1.600		167	70		237	289
0800	3/0		6.60	2.424		375	106		481	570
1000	4/0		5.70	2.807		470	123		593	700
1200	250 kcmil	3 Elec	7.20	3.333		555	146		701	830
1800	Water pipe ground clamps, heavy duty									
2000	Bronze, 1/2" to 1" diameter	1 Elec	8	1	Ea.	21	44		65	88.50
2100	1-1/4" to 2" diameter		8	1		27.50	44		71.50	96
2200	2-1/2" to 3" diameter		6	1.333		46.50	58.50		105	138
2800	Brazed connections, #6 wire		12	.667		12.90	29.50		42.40	57.50
3000	#2 wire		10	.800		17.30	35		52.30	71.50
3100	3/0 wire		8	1		26	44		70	94
3200	4/0 wire		7	1.143		29.50	50		79.50	107
3400	250 kcmil wire		5	1.600		34.50	70		104.50	143
3600	500 kcmil wire		4	2		42.50	88		130.50	178

26 05 33 – Raceway and Boxes for Electrical Systems

26 05 33.05 Conduit

26 05 33.05 Conduit	Crew	Daily Output	Labor-Hours	Unit	Material	2007 Bare Costs Labor	Equipment	Total	Total Incl O&P
0010	CONDUIT To 15' high, includes 2 terminations, 2 elbows,	R260533-22								
0020	11 beam clamps, and 11 couplings per 100 L.F.									
0300	Aluminum, 1/2" diameter	1 Elec	100	.080	L.F.	2.08	3.51		5.59	7.55
0500	3/4" diameter		90	.089		2.81	3.90		6.71	8.90
0700	1" diameter		80	.100		3.72	4.39		8.11	10.65
1000	1-1/4" diameter		70	.114		5	5		10	12.95
1030	1-1/2" diameter		65	.123		6	5.40		11.40	14.65
1050	2" diameter		60	.133		8.20	5.85		14.05	17.70
1070	2-1/2" diameter		50	.160		13.30	7		20.30	25
1100	3" diameter	2 Elec	90	.178		18.15	7.80		25.95	31.50
1130	3-1/2" diameter		80	.200		24	8.80		32.80	39.50
1140	4" diameter		70	.229		29	10.05		39.05	47
1750	Rigid galvanized steel, 1/2" diameter	1 Elec	90	.089		2.48	3.90		6.38	8.55
1770	3/4" diameter		80	.100		2.83	4.39		7.22	9.65
1800	1" diameter		65	.123		3.89	5.40		9.29	12.35
1830	1-1/4" diameter		60	.133		5.40	5.85		11.25	14.65
1850	1-1/2" diameter		55	.145		6.25	6.40		12.65	16.35
1870	2" diameter		45	.178		8	7.80		15.80	20.50
1900	2-1/2" diameter		35	.229		15.05	10.05		25.10	31.50
1930	3" diameter	2 Elec	50	.320		18.40	14.05		32.45	41
1950	3-1/2" diameter		44	.364		23	15.95		38.95	49
1970	4" diameter		40	.400		26	17.55		43.55	54.50
2500	Steel, intermediate conduit (IMC), 1/2" diameter	1 Elec	100	.080		1.92	3.51		5.43	7.35
2530	3/4" diameter		90	.089		2.37	3.90		6.27	8.40
2550	1" diameter		70	.114		3.32	5		8.32	11.10
2570	1-1/4" diameter		65	.123		4.42	5.40		9.82	12.90
2600	1-1/2" diameter		60	.133		5.20	5.85		11.05	14.40
2630	2" diameter		50	.160		6.75	7		13.75	17.85
2650	2-1/2" diameter		40	.200		12.80	8.80		21.60	27
2670	3" diameter	2 Elec	60	.267		16.80	11.70		28.50	36
2700	3-1/2" diameter		54	.296		21	13		34	42.50
2730	4" diameter		50	.320		24	14.05		38.05	47.50
5000	Electric metallic tubing (EMT), 1/2" diameter	1 Elec	170	.047		.60	2.07		2.67	3.74
5020	3/4" diameter	CN	130	.062		.99	2.70		3.69	5.10

Figure 9.110

Credit: *Means Building Construction Cost Data 2007*

While electrical work should be accurately estimated by a subcontractor for bidding purposes, the estimator for the general contractor should understand the work involved. Complete plans and specifications will include a number of schedules and diagrams that will help the estimator to determine quantities. Figures 9.111 and 9.112 are the lighting fixture and panelboard schedules, respectively, for the building project. While such schedules may not always provide quantities, they can serve as a checklist to ensure that all types are counted.

Figure 9.113 is the quantity sheet for lighting. Note that associated boxes, devices, and wiring are taken off at the same time as the fixtures, and that the takeoff is done by floor. By performing the estimate in this way, quantities for purchase are easily determined and changes can be calculated quickly. For example, in a typical project, lighting fixtures are often changed or substituted, either because of aesthetic decisions or unavailability. If quantities and costs are isolated, adjustments are easily made.

The estimate sheets for lighting—as might be prepared by an electrical subcontractor—are shown in Figures 9.114 and 9.115. Prices are from *Means Electrical Cost Data*. By including the associated costs for boxes, devices, and wiring with the lighting fixtures and the other major components, the electrical estimator is able to develop system costs for comparison to other projects, and for cross-checking. Since this is only part of the electrical work, overhead and profit are not likely to be added until the end of the electrical estimate. However, if a breakdown of costs is provided as part of the quotation, each portion of that breakdown would include overhead and profit. Such is the case in the estimate sheet for Divisions 26–28, as prepared by the general contractor, in Figure 9.116. For each system, the associated costs for devices and wiring are included, in addition to overhead and profit.

crushed stone and trenching for drainage may be needed. Consideration must be given to the installation of utilities; this procedure may require removal and reconstruction of the temporary road.

As with the building layout, costs must also be included for the layout and grades of roads, curbs, and associated utilities. Most specifications will state the requirements for work above rough grade: gravel base, subbase, finish paving courses, etc. Bulk cuts and fills for the roadwork should be determined as part of bulk excavation. Excavation for curbs must also be included. Handwork may be necessary.

Bituminous paving is usually taken off and priced by the square yard, but it is sold by the ton. The average weight of bituminous concrete is 145 pounds per cubic foot. By converting, 1 square yard, 1" thick weighs approximately 110 pounds. This figure can be used to determine quantities (in tons) for purchase:

Base course: 2" thick, 1,500 SY

$$1{,}500 \text{ SY} \times 2" \times 110 \text{ lbs} = 330{,}000 \text{ lbs} = 165 \text{ tons}$$

Finish course: 1" thick, 1,500 SY

$$1{,}500 \text{ SY} \times 1" \times 110 \text{ lbs} = 165{,}000 \text{ lbs} = 82.5 \text{ tons}$$

Landscaping

If the required topsoil is stockpiled on site, equipment will be needed for hauling and spreading. Likewise, if the stockpile is too large or too small, trucks will have to be mobilized for transport. If large trees and shrubs are specified, equipment may be required for digging and placing.

Landscaping, including lawn sprinklers, sod, seeding, etc., is usually done by subcontractors. The estimator should make sure that the subcontract includes a certain maintenance period during which all plantings are guaranteed. The required period will be most likely stated in the specifications and should include routine maintenance and replacement of dead plantings.

Estimating Hazardous Waste Remediation/ Cleanup Work

Hazardous waste cleanup/site remediation work is more difficult to estimate than standard construction. The difficulty arises because all projects are unique and the hazardous materials are very different. In addition, the estimator must consider the cost of not only complying with applicable OSHA and other government regulations, but also the reduced worker productivity associated with these regulations. The cost factors associated with compliance include the following:

- Safety training for management and construction personnel
- Pre-employment and conclusion of the work medical exams
- Site safety meetings

- Shower facility and change room
- Worker and equipment decontamination
- Disposal of contaminated water
- Suit up/off time
- Personnel protective equipment
- Nonproductive support workers for crew assistance and safety-related monitoring
- Work breaks due to varying ambient temperature levels

Productivity

The cost impact of these factors is dependent on the types of hazardous waste and the remediation procedure. OSHA regulations and local authorities require different levels of personnel protective gear for different wastes. The levels of protection range from A to D. Level A protection, used when exposed to the most hazardous chemicals, includes a vapor protective suit with air pack, while level D consists of standard construction clothing and safety boots with the possible addition of a Tyvek suit, face mask, or safety glasses. The productivity from workers varies dramatically depending on the level of protective gear when compared to normal construction productivity. In general, D-level protection allows a worker to be approximately 90% productive. Therefore, published construction data can be used to estimate the work.

The productivity of a C-level worker with a full face mask and Tyvek suit is cut to approximately 75%. The B-level protective gear of a full liquid splash-protective suit and air supplied from an outside source drops productivity to approximately 35% efficiency. Level A protection drops worker productivity to 20%. As the level of protection increases, additional nonproductive personnel are also required at the site to monitor and ensure worker safety.

Disposal

Site remediation may consist of on-site encapsulation, and off-site disposal or destruction of the hazardous material by incineration or using an exotic treatment technique. Off-site disposal requires a complete chemical analysis before the material can be moved, acceptance of the material by the proposed recipient, and EPA manifesting (proper shipping documentation). *Note: It is illegal to transport hazardous materials without the proper manifest.*

When estimating off-site disposal, include all transportation costs, waiting time, and disposal fees. When estimating on-site incineration, the estimate must include the cost of trial burns to ensure destruction of the hazardous material and acceptance by the EPA of the stack emissions. The cost of treatment/disposal methods should be determined in conjunction with the manufacturer of the remediation equipment. Hazardous waste disposal is generally specified on a performance basis. Therefore, the process should be prequalified to ensure compliance with the specification.

Summary

Site remediation and hazardous waste cleanup should not be undertaken without complete knowledge of all applicable federal and state regulations. Those companies in the business have compliance departments because meeting all regulations is such an overriding concern in their work. The penalties for noncompliance are severe.

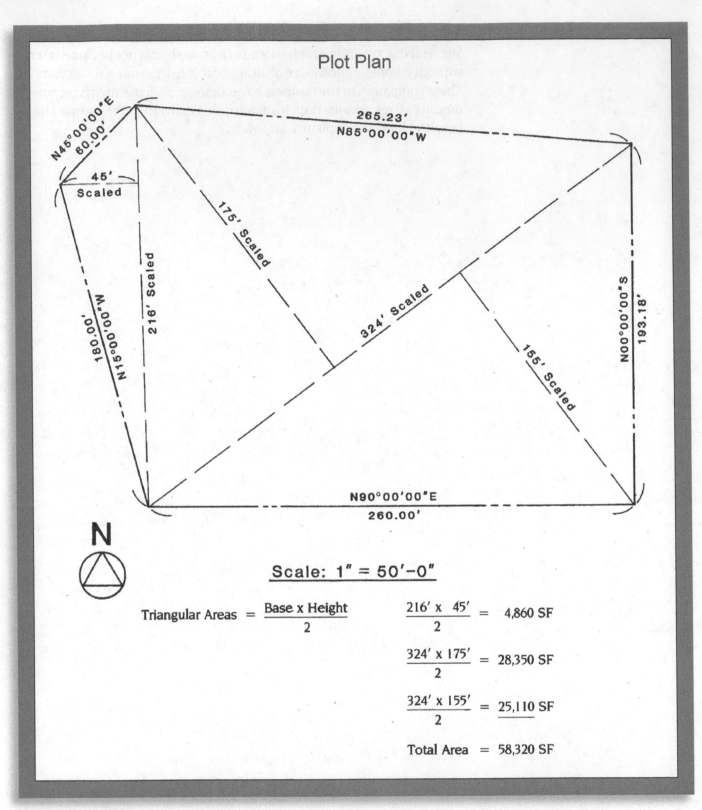

Plot Plan

N45°00'00"E 60.00'

45' Scaled

265.23'
N85°00'00"W

175' Scaled

216' Scaled

324' Scaled

155' Scaled

N00°00'00"S 193.18'

180.00'
N15°00'00"W

N90°00'00"E
260.00'

N

Scale: 1" = 50'-0"

$$\text{Triangular Areas} = \frac{\text{Base x Height}}{2}$$

$$\frac{216' \times 45'}{2} = 4,860 \text{ SF}$$

$$\frac{324' \times 175'}{2} = 28,350 \text{ SF}$$

$$\frac{324' \times 155'}{2} = 25,110 \text{ SF}$$

Total Area = 58,320 SF

Figure 9.117

Level Cross-Section

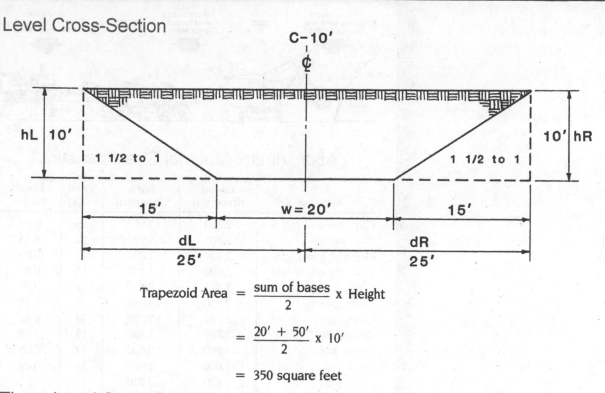

$$\text{Trapezoid Area} = \frac{\text{sum of bases}}{2} \times \text{Height}$$

$$= \frac{20' + 50'}{2} \times 10'$$

$$= 350 \text{ square feet}$$

Three Level Cross Section

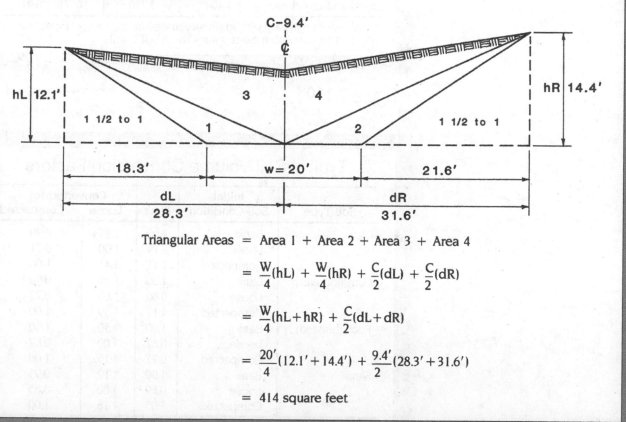

$$\text{Triangular Areas} = \text{Area } 1 + \text{Area } 2 + \text{Area } 3 + \text{Area } 4$$

$$= \frac{W}{4}(hL) + \frac{W}{4}(hR) + \frac{C}{2}(dL) + \frac{C}{2}(dR)$$

$$= \frac{W}{4}(hL + hR) + \frac{C}{2}(dL + dR)$$

$$= \frac{20'}{4}(12.1' + 14.4') + \frac{9.4'}{2}(28.3' + 31.6')$$

$$= 414 \text{ square feet}$$

Figure 9.118

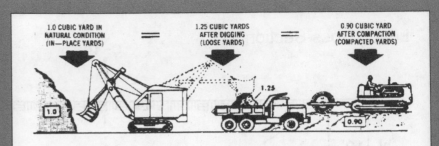

Approximate Material Characteristics*

Material	Loose (lb/cu yd)	Bank (lb/cu yd)	Swell (%)	Load Factor
Clay, dry	2,100	2,650	26	0.79
Clay, wet	2,700	3,575	32	0.76
Clay and gravel, dry	2,400	2,800	17	0.85
Clay and gravel, wet	2,600	3,100	17	0.85
Earth, dry	2,215	2,850	29	0.78
Earth, moist	2,410	3,080	28	0.78
Earth, wet	2,750	3,380	23	0.81
Gravel, dry	2,780	3,140	13	0.88
Gravel, wet	3,090	3,620	17	0.85
Sand, dry	2,600	2,920	12	0.89
Sand, wet	3,100	3,520	13	0.88
Sand and gravel, dry	2,900	3,250	12	0.89
Sand and gravel, wet	3,400	3,750	10	0.91

*Exact values will vary with grain size, moisture content, compaction, etc. Test to determine exact values for specific soils.

Figure 9.119

Credit: *Means Heavy Construction Handbook*

Typical Soil Volume Conversion Factors

Soil Type	Initial Soil Condition	Bank	Converted to: Loose	Converted to: Compacted
Clay	Bank	1.00	1.27	0.90
	Loose	0.79	1.00	0.71
	Compacted	1.11	1.41	1.00
Common earth	Bank	1.00	1.25	0.90
	Loose	0.80	1.00	0.72
	Compacted	1.11	1.39	1.00
Rock (blasted)	Bank	1.00	1.50	1.30
	Loose	0.67	1.00	0.87
	Compacted	0.77	1.15	1.00
Sand	Bank	1.00	1.12	0.95
	Loose	0.89	1.00	0.85
	Compacted	1.05	1.18	1.00

Figure 9.120

Credit: *Means Heavy Construction Handbook*

31 06 Schedules for Earthwork

31 06 60 – Schedules for Special Foundations and Load Bearing Elements

31 06 60.14 Piling Special Costs	Crew	Daily Output	Labor-Hours	Unit	Material	2007 Bare Costs Labor	Equipment	Total	Total Incl O&P
1700 Barge mounted driving rig, add								30%	30%

31 06 60.15 Mobilization

		Crew	Daily Output	Labor-Hours	Unit	Material	Labor	Equipment	Total	Total Incl O&P
0010	**MOBILIZATION**									
0020	Set up & remove, air compressor, 600 C.F.M.	A-5	3.30	5.455	Ea.		157	10.70	167.70	256
0100	1200 C.F.M.	"	2.20	8.182			235	16.05	251.05	385
0200	Crane, with pile leads and pile hammer, 75 ton	B-19	.60	106			3,925	2,750	6,675	9,275
0300	150 ton	"	.36	177			6,550	4,575	11,125	15,500
0500	Drill rig, for caissons, to 36", minimum	B-43	2	24			765	1,750	2,515	3,100
0600	Up to 84"	"	1	48			1,525	3,500	5,025	6,200
0800	Auxiliary boiler, for steam small	A-5	1.66	10.843			310	21.50	331.50	510
0900	Large	"	.83	21.687			625	42.50	667.50	1,025
1100	Rule of thumb: complete pile driving set up, small	B-19	.45	142			5,250	3,675	8,925	12,400
1200	Large	"	.27	237			8,750	6,100	14,850	20,600

31 11 Clearing and Grubbing

31 11 10 – Clearing and Grubbing

31 11 10.10 Clear and Grub

		Crew	Daily Output	Labor-Hours	Unit	Material	Labor	Equipment	Total	Total Incl O&P
0010	**CLEAR AND GRUB**									
0020	Cut & chip light trees to 6" diam.	B-7	1	48	Acre		1,475	1,075	2,550	3,475
0150	Grub stumps and remove	B-30	2	12			390	920	1,310	1,600
0200	Cut & chip medium, trees to 12" diam.	B-7	.70	68.571			2,100	1,550	3,650	4,950
0250	Grub stumps and remove	B-30	1	24			780	1,825	2,605	3,225
0300	Cut & chip heavy, trees to 24" diam.	B-7	.30	160			4,900	3,625	8,525	11,600
0350	Grub stumps and remove	B-30	.50	48			1,550	3,675	5,225	6,425
0400	If burning is allowed, reduce cut & chip									40%
3000	Chipping stumps, to 18" deep, 12" diam.	B-86	20	.400	Ea.		15.35	5.30	20.65	29
3040	18" diameter		16	.500			19.20	6.65	25.85	36.50
3080	24" diameter		14	.571			22	7.60	29.60	41.50
3100	30" diameter		12	.667			25.50	8.85	34.35	48.50
3120	36" diameter		10	.800			30.50	10.65	41.15	58
3160	48" diameter		8	1			38.50	13.30	51.80	72.50
5000	Tree thinning, feller buncher, conifer									
5080	Up to 8" diameter	B-93	240	.033	Ea.		1.28	2.14	3.42	4.28
5120	12" diameter		160	.050			1.92	3.21	5.13	6.40
5240	Hardwood, up to 4" diameter		240	.033			1.28	2.14	3.42	4.28
5280	8" diameter		180	.044			1.71	2.85	4.56	5.70
5320	12" diameter		120	.067			2.56	4.28	6.84	8.55
7000	Tree removal, congested area, aerial lift truck									
7040	8" diameter	B-85	7	5.714	Ea.		176	105	281	390
7080	12" diameter		6	6.667			206	123	329	450
7120	18" diameter		5	8			247	148	395	540
7160	24" diameter		4	10			310	185	495	680
7240	36" diameter		3	13.333			410	246	656	905
7280	48" diameter		2	20			615	370	985	1,350

Figure 9.121

Credit: Means Building Construction Cost Data 2007

Sample Estimate: Divisions 30–39

Site work quantities may be derived from several different drawings. In this case, data is taken from the topographic survey, the finish grading plan, and the site plan (Figures 9.2, 9.3, and 9.4). For smaller projects, these three plans may be combined as one. The detailed information that would normally be included in a complete set of plans and specifications is not provided for this example. For purposes of this demonstration, quantities as shown reflect realistic conditions and are based on normal construction techniques and materials.

For the site clearing portion of the estimate, areas to be cleared are measured from the topographic survey (Figure 9.2), but verification of area and determination of vegetation type must be done during the site visit. The estimate sheets for Divisions 30–39 are shown in Figures 9.122 through 9.125. Note that the units (acres and square yards) used for tree cutting and grubbing are different than those used for light clearing. The estimator should be sure to convert the required units for pricing. For the example, the whole site is to be cleared. The site dimensions are 310' x 365', or 113,150 square feet. Light clearing is priced by the acre, so the area is converted from 12,570 square yards to 2.6 acres. Prices for site clearing are taken from Figure 9.126.

Quantities for bulk excavation are derived using the alternative method as described above and as shown in Figure 9.127. In this example, a grid representing 20' x 20' squares is used. The estimator determines the average elevation change in each square and designates each as a cut (–) or a fill (+). The darkened square within the building is for excavation at the elevator pit. All cuts and fills are totaled separately and converted to cubic yards. An amount to compensate for reduction in volume is added to the total fill yardage before net calculation. This is due to the decrease in volume of earth when compacted. *(See Figure 9.119.)* The costs for bulk excavation are taken from Figure 9.128. Compaction is not included and must be added separately.

Quantities for bulk excavation are based on finish grades. Excess excavation at the foundation to be backfilled must be included as shown in Figure 9.122. Calculations for compaction at the foundation and for loading and hauling excess fill are as follows:

Compaction at foundation:

Excavation (Bank CY) $= 272$ CY

$\dfrac{1.0 \text{ Bank CY} \times 272 \text{ CY}}{0.9 \text{ Compacted CY}} = 302$ CY required

Load and haul excess fill:

Excess fill (Bank CY) $= 2{,}570$ CY

$\dfrac{1.25 \text{ Loose CY} \times 2{,}570 \text{ CY}}{1.0 \text{ Bank CY}} = 3{,}212$ Loose CY

The ratios of bank to loose to compacted earth were derived from Figure 9.120. If the preceding calculations were not made, almost 700 cubic yards of loading and hauling may not have been included in the estimate. This represents 60–70 truckloads. The calculation for foundation compaction indicates that 302 cubic yards of bank material *when compacted* are required to replace 272 cubic yards of excavated bank material. Only backfill (and not compaction) is included for the footings, because the costs are primarily for the spreading of excess material. Compaction is included with the foundation backfill and at the floor subbase.

Costs for asphaltic concrete pavement are taken from Figure 9.129. Note on the estimate sheet that the price per ton for the sample building (Figure 9.124) is different than that in *Means Building Construction Cost Data*. By simple conversion, prices in *Means Building Construction Cost Data* can be adjusted to local conditions:

Line No.: 32 12 16.13 0810	Binder Course
National average	$48.00
Local cost (for sample project)	$53.80

$$\frac{\$53.80}{\$48.00} = 1.12$$

Material cost per SY × Conversion factor = Local cost per SY

$7.85 per SY × 1.12 = $8.79 per SY

The landscaping and site improvements for the sample project are to be subcontracted. This relieves the general contractor not only of the responsibility of pricing, but also of the problems of maintenance and potential plant replacement. The subcontract price, as received by telephone, is shown in Figure 9.130. The estimator must ask the appropriate questions to be sure that all work is included and all requirements are met.

Estimate Summary

At this point in the estimating process, the estimate is complete for Divisions 3 through 33. All vendor quotations and subcontractors' bids should be in hand (although this is not necessarily realistic), and all costs should be determined for the work to be done "in-house." All costs known at this time should be entered on the Estimate Summary sheet as shown in Figure 9.131. Based on the subtotal of these costs, the items in Division 1 that are dependent on job costs can then be calculated.

The Project Overhead Summary is shown in Figure 9.132. In addition to those items related to job costs, there are also certain project overhead items that depend on job duration. A preliminary schedule is required in order to calculate these time-related costs. Using the procedure described in Chapter 5, a preliminary precedence schedule is prepared for the project as shown in Figure 9.133. Note that only those items that affect the project duration are calculated and included in the total time (247 working days). A more extensive and detailed project schedule is completed when the construction contract is awarded. The corresponding preliminary bar schedule is shown in Figure 9.134.

Throughout the estimate for all divisions, the estimator should note all items that should be included as project overhead, as well as those that may affect the project schedule. The estimator must be especially aware of items that are implied but may not be directly stated in the specifications. Some of the "unspecified" costs that may be included are for the preparation of shop drawings, as-built drawings, and the final project schedule. Costs for general job safety—such as railings, safety nets, fire extinguishers, etc.— must also be included if these items are required. If costs for job safety are not listed in the estimate, the items may not be installed on the job—until an OSHA fine, or worse, an accident occurs.

Costs for shop drawings are mentioned above because the time involved in their preparation, submission, and approval must also be considered when preparing the schedule. Construction delays are often blamed on the fact that shop drawings, or "cuts," weren't submitted early enough or approved in a timely fashion. The estimator must base such scheduling decisions on experience.

At this point, all divisions for the building project have been estimated. The preliminary schedule and project overhead summary are complete. Finally, the Estimate Summary sheet can be completed and the final number can be derived. The completed Estimate Summary sheet for the sample estimate is shown in Figure 9.135. Note that the difference between the subtotal of bare costs for all divisions and the "Total Bid" figure is roughly $1,297,000— almost 20%. This amount is the sum of all the indirect costs. *(See Chapters 4 and 7.)*

Sales tax (in this example, 5%) is added to the total bare costs of materials. An additional 10%—for handling and supervision—is added to the material (including tax), equipment, and subcontract costs. This is standard industry practice. Bare labor costs, however, are treated differently. As discussed in Chapter 7, office overhead expenses may be more directly attributable to labor. Consequently, the percentage markup for labor (55.6%) includes employer-paid taxes and insurance, office overhead, and profit. This relationship is shown in Figure 9.136. (Profit may or may not be listed as a separate item.) Since all trades are included in the bare labor subtotal, the percentage used is the average for all skilled workers. The requirements for bonds are determined from the specifications, and their costs are obtained from local bonding agencies. Finally, a contingency is added if the job warrants. This decision is made when the estimator's role comes to an end and the job of the bidder begins. The decision to add a contingency is often based on the relative detail of the plans and specifications, experience, knowledge of the market and competition, and most likely—a gut feeling. The estimate total, with all indirect costs included, may still not be the same number as that which is *finally* submitted as the bid.

Bid day in a contractor's office is usually a hectic time. Many subcontractors delay telephoning, faxing, and e-mailing quotations until the last possible moment in order to prevent the possibility of bid shopping. Subs to subcontractors also withhold quotes in the same way and for the same reason. This situation, in many instances, forces the prime contractor to use a budget, or even a "ballpark," figure so that the bid sheets can be tabulated to obtain a final number. The importance of the estimator's reviewing all disciplines of the estimate cannot be overstressed. The chaotic atmosphere of bid day can only be compensated by thoroughness, precision, and attention to detail during the whole preceding estimating process. If an estimate is poorly prepared and put together, flaws will be amplified at these late stages.

Many general or prime contractors or construction managers use an adjustment column to arrive at a "quote" in the last hour before a bid is due. Such adjustments might appear as follows:

Structural Steel	+ $10,000
Mechanical	− 35,000
Partitions	− 10,000
Electrical (Use 2nd Bidder)	+ 6,000
Total Deduct	$29,000

These deductions or additions may be obtained from last minute phone calls, faxed quotes, error corrections, or discovered omissions. The final number, or bid price, is, in many cases, arrived at by a principal or senior member of the firm. Bidding strategies are discussed in Chapter 6, but the final quotation is usually determined by personal judgment, good or bad.

Factors that may affect a final decision are: the risk involved, competition from other bidders, thoroughness of the plans and specs, and above all, the years of experience—the qualification to make such a judgment. Some firms are so scientific and calculating that every quoted price includes a "lucky number."

A great deal of success in bidding can be attributed to the proper choice of jobs to bid. A contracting firm can go broke estimating every available job. The company must be able to recognize which jobs are too risky and when the competition is too keen, while not overlooking those that can be profitable. Again, knowledge of the marketplace and experience are the keys to successful bidding.

Conclusion

The primary purpose of this text has not been to tell the reader how much an item will cost, but instead, how to develop a consistent and thorough approach to the estimating process. If such a pattern is developed—employing consistency, attention to detail, experience, and above all, common sense— accurate costs will follow.

If an estimate is thorough, organized, neat, and concise, the benefits go beyond winning contracts. The information and data that is developed will be useful throughout a project—for purchasing, change orders, cost accounting and control, and development of historical costs.

CONDENSED ESTIMATE SUMMARY

PROJECT	Office Building					SHEET NO.	
LOCATION		TOTAL AREA / VOLUME				ESTIMATE NO:	
ARCHITECT	As Shown	COST PER S.F. / C.F.				DATE: Jan/2007	
PRICES BY: DEF		EXTENSIONS BY: DEF				NO. OF STORIES	
						CHECKED BY: GHI	

DIV.	DESCRIPTION	Mat'l	Labor	Equip.	Sub w/ Tax		Total
1	General Requirements						
3	Concrete	$424,530	$285,687	$8,658			
4	Masonry	$39,133	$35,675	$7,366			
5	Metals				$929,947		
6	Wood, Plastics, and Composites	$4,427	$2,856				
7	Thermal & Moisture Protection	$14,481	$8,418		$111,691		
8	Openings	$36,965	$6,218		$1,176,869		
9	Finishes	$161,920	$104,662		$407,808		
10	Specialties	$18,489	$2,925				
11	Equipment						
12	Furnishings	$2,921	$627				
13	Special Construction						
14	Conveying Systems				$204,710		
21, 22 & 23	Fire suppression, Plumbing, and Heating Ventilating and Air Conditioning				$1,082,187		
26, 27 & 28	Electrical, Communication, and Electronic Safety & Security				$814,827		
31, 32 33	Site Construction - Earthwork, Exterior Improvements & Utilities	$245,557	$53,319	$41,332	$51,872		
	Subtotals						
	Sales Tax - Mat. & Equip. 5% Subcontract 2.5%						
	Overhead & Profit 10%M, 55.6%L 10%E, 10%S						
	Subtotal						
	Bond 1.2%						
	Contingency 2%						
	Adjustments						
	TOTAL BID						

Figure 9.131

To download this and other forms in this book, visit www.rsmeans.com/supplement/67303B.asp

PROJECT
OVERHEAD SUMMARY

PROJECT: Office Building

LOCATION:

QUANTITIES BY: ABC PRICES BY: DEF

ARCHITECT:

EXTENSIONS BY: DEF

SHEET NO. 1 of 2

ESTIMATE NO:

DATE: Jan-07

CHECKED BY: GHI

DESCRIPTION	QUANTITY	UNIT	MATERIAL/EQUIP. UNIT	MATERIAL/EQUIP. TOTAL	LABOR UNIT	LABOR TOTAL	TOTAL COST UNIT	TOTAL COST TOTAL
Job Organization: Superintendent	49	Week			1650	80850		
Project Manager								
Timekeeper & Material Clerk	40	Week			350	14000		
Clerical								
Safety, Watchman & First Aid								
Travel Expense: Superintendent								
Project Manager								
Engineering: Layout (3 person crew)	10	Day			1085	10850		
Inspection / Quantities								
Drawings								
CPM Schedule								
Testing: Soil	1	LS		9950				
Materials								
Structural								
Equipment: Cranes								
Concrete Pump, Conveyor, Etc.								
Elevators, Hoists								
Freight & Hauling								
Loading, Unloading, Erecting, Etc.								
Maintenance								
Pumping								
Scaffolding								
Small Power Equipment / Tools								
Field Offices: Job Office, Trailer	11	Mo	330	3630				
Architect / Owner's Office								
Temporary Telephones	11	Mo	210	2310				
Utilities								
Temporary Toilets	11	Mo	185	2035				
Storage Areas & Sheds								
Temporary Utilities: Heat								
Light & Power	567	CSF	16.05	9100				
PAGE TOTALS			$	27,025	$	105,700		

Figure 9.132

DESCRIPTION	QUANTITY	UNIT	MATERIAL/EQUIP.		LABOR		TOTAL COST	
			UNIT	TOTAL	UNIT	TOTAL	UNIT	TOTAL
Totals Brought Forward				$27,025		$105,700		
Winter Protection: Temp. Heat/Protection	56700	SF	0.39	22113	0.61	34587		
Snow Plowing								
Thawing Materials								
Temporary Roads	750	SY	4.21	3158	2.04	1530		
Signs & Barricades: Site Sign	1	LS	160	160				
Temporary Fences	1350	LF	6.50	8775	1.53	2066		
Temporary Stairs, Ladders & Floors								
Photographs								
Clean Up								
Dumpster	40	Week	690	27600				
Final Clean Up	56.7	MSF	5.78	328	45	2552		
Continuous - One Laborer	45	Week			1150	51750		
Punch List	0.2	%		12200				
Permits: Building	1	%		63000				
Misc.								
Insurance: Builders Risk - Additional Rider	1	%		63000				
Owner's Protective Liability								
Umbrella								
Unemployment Ins. & Social Security								
Bonds								
Performance								
Material & Equipment								
Main Office Expense								
Special Items								
Totals:				$227,359		$198,184		

Figure 9.132 (cont.)

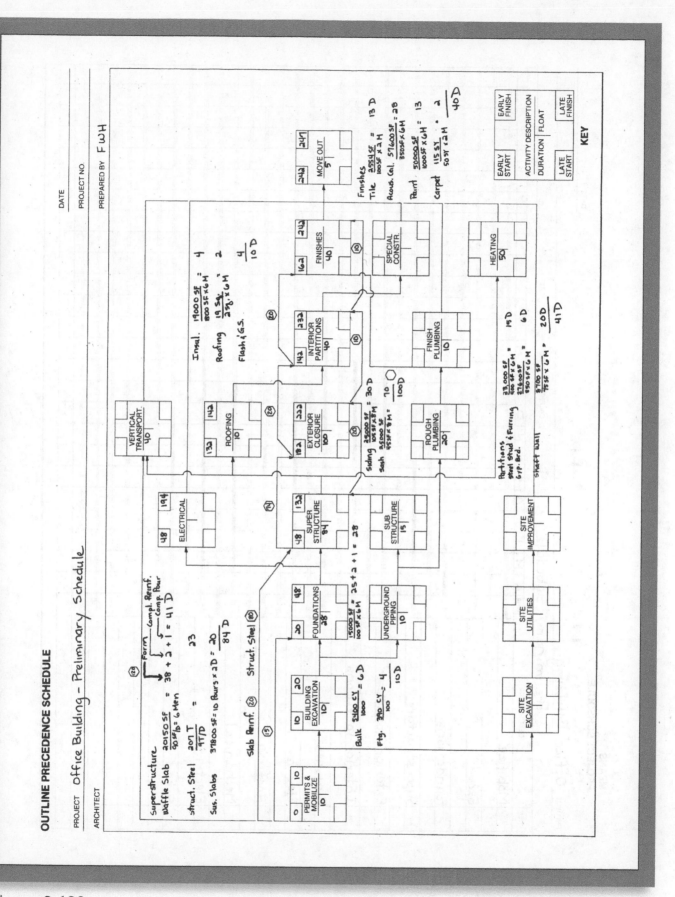

Figure 9.133

339

Installation Time in Labor-Hours for Block Walls, Partitions, and Accessories

Description	Labor-Hours	Unit
Foundation Walls, Trowel Cut Joints, Parged		
1/2" Thick, 1 Side, 8" x 16" Face		
Hollow		
8" Thick	.093	S.F.
12" Thick	.122	S.F.
Solid		
8" Thick	.096	S.F.
12" Thick	.126	S.F.
Backup Walls, Tooled Joint 1 Side,		
8" x 16" Face		
4" Thick	.091	S.F.
8" Thick	.100	S.F.
Partition Walls, Tooled Joint 2 Sides		
8" x 16" Face		
Hollow		
4" Thick	.093	S.F.
8" Thick	.107	S.F.
12" Thick	.141	S.F.
Solid		
4" Thick	.096	S.F.
8" Thick	.111	S.F.
12" Thick	.148	S.F.
Stud Block Walls, Tooled Joints 2 Sides		
8" x 16" Face		
6" Thick and 2", Plain	.098	S.F.
Embossed	.103	S.F.
10" Thick and 2", Plain	.108	S.F.
Embossed	.114	S.F.
6" Thick and 2" Each Side, Plain	.114	S.F.
Acoustical Slotted Block Walls		
Tooled 2 Sides		
4" Thick	.127	S.F.
8" Thick	.151	S.F.
Glazed Block Walls, Tooled Joint 2 Sides		
8" x 16", Glazed 1 Face		
4" Thick	.116	S.F.
8" Thick	.129	S.F.
12" Thick	.171	S.F.
8" x 16", Glazed 2 Faces		
4" Thick	.129	S.F.
8" Thick	.148	S.F.
8" x 16", Corner		
4" Thick	.140	Ea.

(continued on next page)

Installation Time in Labor-Hours for Block Walls, Partitions, and Accessories (continued)

Description	Labor-Hours	Unit
Structural Facing Tile, Tooled 2 Sides		
5" x 12", Glazed 1 Face		
4" Thick	.182	S.F.
8" Thick	.222	S.F.
5" x 12", Glazed 2 Faces		
4" Thick	.205	S.F.
8" Thick	.246	S.F.
8" x 16", Glazed 1 Face		
4" Thick	.116	S.F.
8" Thick	.129	S.F.
8" x 16", Glazed 2 Faces		
4" Thick	.123	S.F.
8" Thick	.137	S.F.
Exterior Walls, Tooled Joint 2 Sides, Insulated		
8" x 16" Face, Regular Weight		
8" Thick	.110	S.F.
12" Thick	.145	S.F.
Lightweight		
8" Thick	.104	S.F.
12" Thick	.137	S.F.
Architectural Block Walls, Tooled Joint 2 Sides		
8" x 16" Face		
4" Thick	.116	S.F.
8" Thick	.138	S.F.
12" Thick	.181	S.F.
Interlocking Block Walls, Fully Grouted		
Vertical Reinforcing		
8" Thick	.131	S.F.
12" Thick	.145	S.F.
16" Thick	.173	S.F.
Bond Beam, Grouted, 2 Horizontal Rebars		
8" x 16" Face, Regular Weight		
8" Thick	.133	L.F.
12" Thick	.192	L.F.
Lightweight		
8" Thick	.131	L.F.
12" Thick	.188	L.F.
Lintels, Grouted, 2 Horizontal Rebars		
8" x 16" Face, 8" Thick	.119	L.F.
16" x 16" Face, 8" Thick	.131	L.F.
Control Joint 4" Wall	.013	L.F.
8" Wall	.020	L.F.
Grouting Bond Beams and Lintels		
8" Deep Pumped, 8" Thick	.018	L.F.
12" Thick	.025	L.F.
Concrete Block Cores Solid		
4" Thick By Hand	.035	S.F.
8" Thick Pumped	.038	S.F.
Cavity Walls 2" Space Pumped	.016	S.F.
6" Space	.034	S.F.

(continued on next page)

Installation Time in Labor-Hours for Block Walls, Partitions, and Accessories (continued)

Description	Labor-Hours	Unit
Joint Reinforcing		
Wire Strips Regular Truss to 6" Wide	.267	C.L.F.
12" Wide	.400	C.L.F.
Cavity Wall with Drip Section to 6" Wide	.267	C.L.F.
12" Wide	.400	C.L.F.
Lintels Steel Angles Minimum	.008	lb.
Maximum	.016	lb.
Wall Ties	.762	C
Coping For 12" Wall Stock Units, Aluminum	.200	L.F.
Precast Concrete	.188	L.F.
Structural Reinforcing, Placed Horizontal,		
#3 and #4 Bars	.018	lb.
#5 and #6 Bars	.010	lb.
Placed Vertical, #3 and #4 Bars	.023	lb.
#5 and #6 Bars	.012	lb.
Acoustical Slotted Block		
4" Thick	.127	S.F.
6" Thick	.138	S.F.
8" Thick	.151	S.F.
12" Thick	.163	S.F.
Lightweight Block		
4" Thick	.090	S.F.
6" Thick	.095	S.F.
8" Thick	.100	S.F.
10" Thick	.103	S.F.
12" Thick	.130	S.F.
Regular Block		
Hollow		
4" Thick	.093	S.F.
6" Thick	.100	S.F.
8" Thick	.107	S.F.
10" Thick	.111	S.F.
12" Thick	.141	S.F.
Solid		
4" Thick	.095	S.F.
6" Thick	.105	S.F.
8" Thick	.113	S.F.
12" Thick	.150	S.F.
Glazed Concrete Block		
Single Face 8" x 16"		
2" Thick	.111	S.F.
4" Thick	.116	S.F.
6" Thick	.121	S.F.
8" Thick	.129	S.F.
12" Thick	.171	S.F.
Double Face		
4" Thick	.129	S.F.
6" Thick	.138	S.F.
8" Thick	.148	S.F.

(continued on next page)

Installation Time in Labor-Hours for Block Walls, Partitions, and Accessories (continued)

Description	Labor-Hours	Unit
Joint Reinforcing Wire Strips		
4" and 6" Wall	.267	C.L.F.
8" Wall	.320	C.L.F.
10" and 12" Wall	.400	C.L.F.
Steel Bars Horizontal		
#3 and #4	.018	lb.
#5 and #6	.010	lb.
Vertical		
#3 and #4	.023	lb.
#5 and #6	.012	lb.
Grout Cores Solid		
By Hand 6" Thick	.035	S.F.
Pumped 8" Thick	.038	S.F.
10" Thick	.039	S.F.
12" Thick	.040	S.F.

Installation Time in Labor-Hours for the Erection of Steel Superstructure Systems

Description	Labor-Hours	Unit
Steel Columns Concrete Filled		
4" Diameter	.072	L.F.
5" Diameter	.055	L.F.
6-5/8" Diameter	.047	L.F.
Steel Pipe		
6" Diameter	6.000	ton
12" Diameter	2.000	ton
Structural Tubing		
6" x 6"	6.000	ton
10" x 10"	2.000	ton
Wide Flange		
W8 x 31	3.355	ton
W10 x 45	2.412	ton
W12 x 50	2.171	ton
W14 x 74	1.538	ton
Beams WF Average	4.000	ton
Steel Joists H Series Horizontal Bridging		
To 30' Span	6.667	ton
30' to 50' Span	6.353	ton
(Includes One Row of Bolted Cross Bridging for Spans Over 40' Where Required)		
LH Series Bolted Cross Bridging		
Spans to 96'	6.154	ton
DLH Series Bolted Cross Bridging		
Spans to 144' Shipped in 2 Pieces	6.154	ton
Joist Girders	6.154	ton
Trusses, Factory Fabricated, with Chords	7.273	ton
Metal Decking Open Type		
1-1/2" Deep		
22 Gauge	.007	S.F.
18 and 20 Gauge	.008	S.F.
3" Deep		
20 and 22 Gauge	.009	S.F.
18 Gauge	.010	S.F.
16 Gauge	.011	S.F.
4-1/2" Deep		
20 Gauge	.012	S.F.
18 Gauge	.013	S.F.
16 Gauge	.014	S.F.
7-1/2" Deep		
18 Gauge	.019	S.F.
16 Gauge	.020	S.F.

(continued on next page)

Installation Time in Labor-Hours for the Erection of Steel Superstructure Systems (continued)

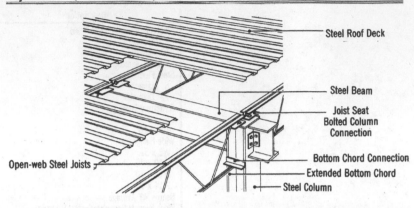

Steel Roof Deck

Steel Beam

Joist Seat
Bolted Column
Connection

Bottom Chord Connection
Extended Bottom Chord
Steel Column

Open-web Steel Joists

Installation Time in Labor-Hours for Fireproofing Structural Steel

Description	Labor-Hours	Unit
Fireproofing - 10" Column Encasements		
Perlite Plaster	.273	V.L.F.
1" Perlite on 3/8" Gypsum Lath	.345	V.L.F.
Sprayed Fiber	.131	V.L.F.
Concrete 1-1/2" Thick	.716	V.L.F.
Gypsum Board 1/2" Fire Resistant,		
1 Layer	.364	V.L.F.
2 Layer	.428	V.L.F.
3 Layer	.530	V.L.F.
Fireproofing – 16" x 7" Beam Encasements		
Perlite Plaster on Metal Lath	.453	L.F.
Gypsum Plaster on Metal Lath	.408	L.F.
Sprayed Fiber	.079	L.F.
Concrete 1-1/2" Thick	.554	L.F.
Gypsum Board 5/8" Fire Resistant	.488	L.F.

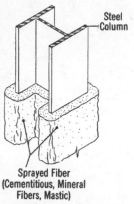

Sprayed Fiber on Columns

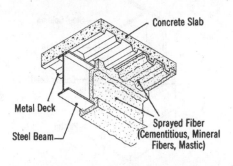

Sprayed Fiber on Beams and Girders

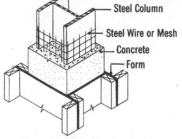

Concrete Encasement on Columns

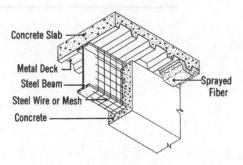

Concrete Encasement on Beams and Girders

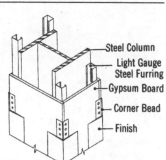

Gypsum Board on Columns

(continued on next page)

Installation Time in Labor-Hours for Fireproofing Structural Steel (continued)

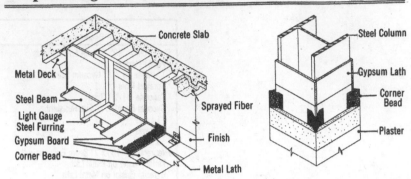

Gypsum Board on Beams and Girders   Plaster on Gypsum Lath — Columns

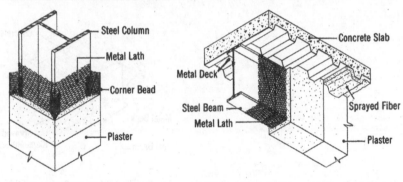

Plaster on Metal Lath — Columns   Plaster on Metal Lath — Beams and Girders

Flat Roof Framing

This table can be used to compute the amount of lumber in board feet required to frame a flat roof, based on the size lumber required.

	Flat Roof Framing		
Joist Size	Inches On Center	Board Feet per Square Foot of Ceiling Area	Nails Lbs. per MBM
2" x 6"	12"	1.17	10
	16"	.91	10
	20"	.76	10
	24"	.65	10
2" x 8"	12"	1.56	8
	16"	1.21	8
	20"	1.01	8
	24"	.86	8
2" x 10"	12"	1.96	6
	16"	1.51	6
	20"	1.27	6
	24"	1.08	6
2" x 12"	12"	2.35	5
	16"	1.82	5
	20"	1.52	5
	24"	1.30	5
3" x 8"	12"	2.35	5
	16"	1.82	5
	20"	1.52	5
	24"	1.30	5
3" x 10"	12"	2.94	4
	16"	2.27	4
	20"	1.90	4
	24"	1.62	4

MBM; MFBM = Thousand Feet Board Measure

Pitched Roof Framing

This table is used to compute the board feet of lumber required per square foot of roof area, based on the required spacing and the lumber size to be used.

	Rafters Including Collar Ties, Hip and Valley Rafters, Ridge Poles							
	Spacing Center to Center							
Rafter Size	12"		16"		20"		24"	
	Board Feet per Square Foot of Roof Area	Nails Lbs. per MBM	Board Feet per Square Foot of Roof Area	Nails Lbs. per MBM	Board Feet per Square Foot of Roof Area	Nails Lbs. per MBM	Board Feet per Square Foot of Roof Area	Nails Lbs. per MBM
2" x 4"	.89	17	.71	17	.59	17	.53	17
2" x 6"	1.29	12	1.02	12	.85	12	.75	12
2" x 8"	1.71	9	1.34	9	1.12	9	.98	9
2" x 10"	2.12	7	1.66	7	1.38	7	1.21	7
2" x 12"	2.52	6	1.97	6	1.64	6	1.43	6
3" x 8"	2.52	6	1.97	6	1.64	6	1.43	6
3" x 10"	3.13	5	2.45	5	2.02	5	1.78	5

Board Feet Required for On-the-Job Cut Bridging

Based on the size of the lumber used for joists, the total lengths of lumber for various sized bridging can be obtained from this chart.

Joist Size	Spacing	1" x 3"		1" x 4"		2" x 3"	
		B.F.	Nails	B.F.	Nails	B.F.	Nails
2" x 8"	12"	.04	147	.05	112	.08	77
	16"	.04	120	.05	91	.08	61
	20"	.04	102	.05	77	.08	52
	24"	.04	83	.05	63	.08	42
2" x 10"	12"	.04	136	.05	103	.08	71
	16"	.04	114	.05	87	.08	58
	20"	.04	98	.05	74	.08	50
	24"	.04	80	.05	61	.08	41
2" x 12"	12"	.04	127	.05	96	.08	67
	16"	.04	108	.05	82	.08	55
	20"	.04	94	.05	71	.08	48
	24"	.04	78	.05	59	.08	39
3" x 8"	12"	.04	160	.05	122	.08	84
	16"	.04	127	.05	96	.08	66
	20"	.04	107	.05	81	.08	54
	24"	.04	86	.05	65	.08	44
3" x 10"	12"	.04	146	.05	111	.08	77
	16"	.04	120	.05	91	.08	62
	20"	.04	102	.05	78	.08	52
	24"	.04	83	.05	63	.08	42

Table title: Cross Bridging—Board Feet per Square Foot of Floors, Ceiling or Flat Roof Area; Nails — Pounds Per MBM of Bridging

Installation Time in Labor-Hours for Interior Doors and Frames

Description	Labor-Hours	Unit
Architectural, Flush, Interior, Hollow Core, Veneer Face		
Up to 3'-0" x 7' x 0"	1.020	Ea.
4'-0" x 7'-0"	1.080	Ea.
High Pressured Plastic Laminate Face		
Up to 2'-6" x 6'-8"	1.000	Ea.
3'-0" x 7'-0"	1.153	Ea.
4'-0" x 7'-0"	1.234	Ea.
Particle Core, Veneer Face		
2'-6" x 6'-8"	1.067	Ea.
3'-0" x 6'-8"	1.143	Ea.
3'-0" x 7'-0"	1.231	Ea.
4'-0" x 7'-0"	1.333	Ea.
M.D.O. on Hardboard Face		
3'-0" x 7'-0"	1.333	Ea.
4'-0" x 7'-0"	1.600	Ea.
High Pressure Plastic Laminate Face		
3'-0" x 7'-0"	1.455	Ea.
4'-0" x 7'-0"	2.000	Ea.
Flush, Exterior, Solid Core, Veneer Face		
2'-6" x 7'-0"	1.067	Ea.
3'-0" x 7'-0"	1.143	Ea.
Decorator, Hand Carved Solid Wood		
Up to 3'-0" x 7'-0"	1.143	Ea.
3'-6" x 7'-0"	1.231	Ea.
Fire Door, Flush, Mineral Core B Label, 1 Hour, Veneer Face		
2'-6" x 6'-8"	1.143	Ea.
3'-0" x 7'-0"	1.333	Ea.
4'-0" x 7'-0"	1.333	Ea.
High Pressure Plastic Laminate Face		
3'-0" x 7'-0"	1.455	Ea.
4'-0" x 7'-0"	1.600	Ea.
Residential, Interior		
Hollow Core or Panel		
Up to 2'-8" x 6'-8"	.889	Ea.
3'-0" x 6'-8"	.941	Ea.
Bi-Folding Closet		
3'-0" x 6'-8"	1.231	Ea.
5'-0" x 6'-8"	1.455	Ea.
Interior Prehung, Hollow Core or Panel		
Up to 2'-8" x 6'-8"	.800	Ea.
3'-0" x 6'-8"	.842	Ea.
Exterior, Entrance, Solid Core or Panel		
Up to 2'-8" x 6'-8"	1.000	Ea.
3'-0" x 6'-8"	1.067	Ea.
Exterior Prehung, Entrance		
Up to 3'-0" x 7'-0"	1.000	Ea.

(continued on next page)

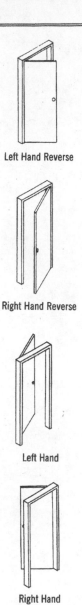

Left Hand Reverse

Right Hand Reverse

Left Hand

Right Hand

Hand Designations

Installation Time in Labor-Hours for Interior Doors and Frames (continued)

Description	Labor-Hours	Unit
Hollow Metal Doors Flush		
Full Panel, Commercial		
20 Gauge		
2'-0" x 6'-8"	.800	Ea.
2'-6" x 6'-8"	.888	Ea.
3'-0" x 6'-8" or 3'-0" x 7'-0"	.941	Ea.
4'-0" x 7'-0"	1.066	Ea.
18 Gauge		
2'-6" x 6'-8" or 2'-6" x 7'-0"	.941	Ea.
3'-0" x 6'-8" or 3'-0" x 7'-0"	1.000	Ea.
4'-0" x 7'-0"	1.066	Ea.
Residential		
24 Gauge		
2'-8" x 6'-8"	1.000	Ea.
3'-0" x 7'-0"	1.066	Ea.
Bifolding		
3'-0" x 6'- 8"	1.000	Ea.
5'-0" x 6'-8"	1.143	Ea.
Steel Frames		
18 Gauge		
3'-0" Wide	1.000	Ea.
6'-0" Wide	1.142	Ea.
16 Gauge		
4'-0" Wide	1.066	Ea.
8'-0" Wide	1.333	Ea.
Transom Lite Frames		
Fixed Add	.103	S.F.
Movable Add	.123	S.F.

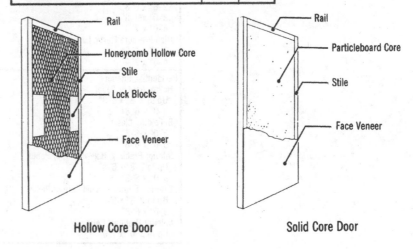

Hollow Core Door

Solid Core Door

Partition Density Guide

This table allows you to estimate the average quantities of partitions found in various types of buildings when no interior plans are available, such as in a conceptual estimating situation. To use this chart, pick out the type of structure to be estimated from the left-most column. Moving across, choose from the second column the number of stories of the building. Read across to the "Partition Density" column. This figure represents the number of square feet of floor area for every linear foot of partition.

Example: The project to be estimated is a three-story office building of 10,000 square feet per floor (total of 30,000 square feet). To estimate the approximate quantity of interior partitions in the entire building, first look in column 1 for "Office." Reading across to the "Stories" column, find the "3-5 Story" line. The "Partition Density" column for that line indicates 20 S.F./L.F. To find the number of partitions, divide the building area by the density factor: 30,000 S.F./20 = 1,500 L.F. of partition. Note that the right-most column gives a breakdown of average partition types. For our office building, the average mix is 30% concrete block (generally found at stairwells and service area partitions) and 70% drywall. Our drywall total would then be 1,050 L.F. To determine the total square footage of partitions, multiply this linear footage by the specified (or assumed) partition height. Note that the remainder of the partitions (450 L.F.) must be added to the masonry portion of the estimate, or at least accounted for in the estimate.

Building Type	Stories	Partition Density	Description of Partition
Apartments	1 story	9 S.F./L.F.	Plaster, wood doors & trim
	2 story	8 S.F./L.F.	Drywall, wood studs, wood doors & trim
	3 story	9 S.F./L.F.	Plaster, wood studs, and wood doors & trim
	5 story	9 S.F./L.F.	Plaster, wood studs, wood doors & trim
	6–15 story	8 S.F./L.F.	Drywall, wood studs, wood doors & trim
Bakery	1 story	50 S.F./L.F.	Conc. block, paint, door & drywall, wood studs
	2 story	50 S.F./L.F.	Conc. block, paint, door & drywall, wood studs
Bank	1 story	20 S.F./L.F.	Plaster, wood studs, wood doors & trim
	2-4 story	15 S.F./L.F.	Plaster, wood studs, wood doors & trim
Bottling Plant	1 story	50 S.F./L.F.	Conc. block, drywall, wood studs, wood trim
Bowling Alley	1 story	50 S.F./L.F.	Conc. block, wood & metal doors, wood trim
Bus Terminal	1 story	15 S.F./L.F.	Conc. block, ceramic tile, wood trim
Cannery	1 story	100 S.F./L.F.	Drywall on metal studs
Car Wash	1 story	18 S.F./L.F.	Concrete block, painted & hollow metal door
Dairy Plant	1 story	30 S.F./L.F.	Concrete block, glazed tile, insulated cooler doors
Department Store	1 story	60 S.F./L.F.	Drywall, wood studs, wood doors & trim
	2-5 story	60 S.F./L.F.	30% concrete block, 70% drywall, wood studs
Dormitory	2 story	9 S.F./L.F.	Plaster, concrete block, wood doors & trim
	3–5 story	9 S.F./L.F.	Plaster, concrete block, wood doors & trim
	6–15 story	9 S.F./L.F.	Plaster, concrete block, wood doors & trim
Funeral Home	1 story	15 S.F./L.F.	Plaster on concrete block & wood studs, paneling
	2 story	14 S.F./L.F.	Plaster, wood studs, paneling & wood doors
Garage Sales & Service	1 story	30 S.F./L.F.	50% conc. block, 50% drywall, wood studs
Hotel	3–8 story	9 S.F./L.F.	Plaster, conc. block, wood doors & trim
	9–15 story	9 S.F./L.F.	Plaster, conc. block, wood doors & trim
Laundromat	1 story	25 S.F./L.F.	Drywall, wood studs, wood doors & trim
Medical Clinic	1 story	6 S.F./L.F.	Drywall, wood studs, wood doors & trim
	2-4 story	6 S.F./L.F.	Drywall, wood studs, wood doors & trim
Motel	1 story	7 S.F./L.F.	Drywall, wood studs, wood doors & trim
	2-3 story	7 S.F./L.F.	Concrete block, drywall on wood studs, wood paneling

(continued on next page)

Partition Density Guide (continued)

Building Type	Stories	Partition Density	Description of Partition
Movie 200–600 seats Theater 601–1400 seats 1401–2200 seats	1 story	18 S.F./L.F. 20 S.F./L.F. 25 S.F./L.F.	Concrete block, wood, metal, vinyl trim Concrete block, wood, metal, vinyl trim Concrete block, wood, metal, vinyl trim
Nursing Home	1 story 2–4 story	8 S.F./L.F. 8 S.F./L.F.	Drywall, wood studs, wood doors & trim Drywall, wood studs, wood doors & trim
Office	1 story 2 story 3–5 story 6–10 story 11–20 story	20 S.F./L.F. 20 S.F./L.F. 20 S.F./L.F. 20 S.F./L.F. 20 S.F./L.F.	30% concrete block, 70% drywall on wood studs 30% concrete block, 70% drywall on wood studs 30% concrete block, 70% movable partitions 30% concrete block, 70% movable partitions 30% concrete block, 70% movable partitions
Parking Ramp (Open) Parking Garage	2–8 story 2–8 story	60 S.F./L.F. 60 S.F./L.F.	Stair and elevator enclosures only Stair and elevator enclosures only
Pre-Engineered Store Office Shop	1 story 1 story 1 story	60 S.F./L.F. 15 S.F./L.F. 15 S.F./L.F.	Drywall on wood studs, wood doors & trim Concrete block, movable wood partitions Movable wood partitions
Radio & TV Broadcasting & TV Transmitter	1 story 1 story	25 S.F./L.F. 40 S.F./L.F.	Concrete block, metal and wood doors Concrete block, metal and wood doors
Self Service Restaurant Cafe & Drive-in Restaurant Restaurant with seating Supper Club Bar or Lounge	1 story 1 story 1 story 1 story 1 story	15 S.F./L.F. 18 S.F./L.F. 25 S.F./L.F. 25 S.F./L.F. 24 S.F./L.F.	Concrete block, wood and aluminum trim Drywall, wood studs, ceramic & plastic trim Concrete block, paneling, wood studs & trim Concrete block, paneling, wood studs & trim Plaster or gypsum lath, wood studs
Retail Store or Shop	1 story	60 S.F./L.F.	Drywall wood studs, wood doors & trim
Service Station Masonry Metal panel Frame	1 story 1 story 1 story	15 S.F./L.F. 15 S.F./L.F. 15 S.F./L.F.	Concrete block, paint, door & drywall, wood studs Concrete block paint door & drywall, wood studs Drywall, wood studs, wood doors & trim
Shopping Center (strip) (group)	1 story 1 story 2 story	30 S.F./L.F. 40 S.F./L.F. 40 S.F./L.F.	Drywall, wood studs, wood doors & trim 50% concrete block, 50% drywall, wood studs 50% concrete block, 50% drywall, wood studs
Small Food Store	1 story	30 S.F./L.F.	Concrete block drywall, wood studs, wood trim
Store/Apt. Masonry above Frame Frame	2 story 2 story 3 story	10 S.F./L.F. 10 S.F./L.F. 10 S.F./L.F.	Plaster, wood studs, wood doors & trim Plaster wood studs, wood doors & trim Plaster, wood studs and wood doors & trim
Supermarkets	1 story	40 S.F./L.F.	Concrete block, paint, drywall & porcelain panel
Truck Terminal	1 story	0	
Warehouse	1 story	0	

Installation Time in Labor-Hours for Wood Stud Partition Systems

Description	Labor-Hours	Unit
Wood Partitions Studs with Single Bottom Plate and Double Top Plate		
2" x 3" or 2" x 4" Studs		
12" On Center	.020	S.F.
16" On Center	.016	S.F.
24" On Center	.013	S.F.
2" x 6" Studs		
12" On Center	.023	S.F.
16" On Center	.018	S.F.
24" On Center	.014	S.F.
Plates		
2" x 3"	.019	L.F.
2" x 4"	.020	L.F.
2" x 6"	.021	L.F.
Studs		
2" x 3"	.013	L.F.
2" x 4"	.012	L.F.
2" x 6"	.016	L.F.
Blocking	.032	L.F.
Grounds 1" x 2"		
For Casework	.024	L.F.
For Plaster	.018	L.F.
Insulation Fiberglass Batts	.005	S.F.
Metal Lath Diamond Expanded		
2.5 lb. per S.Y.	.094	S.Y.
3.4 lb. per S.Y.	.100	S.Y.
Gypsum Lath		
3/8" Thick	.094	S.Y.
1/2" Thick	.100	S.Y.
Gypsum Plaster		
2 Coats	.381	S.Y.
3 Coats	.460	S.Y.
Perlite or Vermiculite Plaster		
2 Coats	.435	S.Y.
3 Coats	.541	S.Y.
Wood Fiber Plaster		
2 Coats	.556	S.Y.
3 Coats	.702	S.Y.

(continued on next page)

Installation Time in Labor-Hours
for Wood Stud Partition Systems (continued)

Description	Labor-Hours	Unit
Drywall Gypsum Plasterboard Including Taping		
3/8" Thick	.015	S.F.
1/2" or 5/8" Thick	.017	S.F.
For Thin Coat Plaster Instead of Taping Add	.013	S.F.
Prefinished Vinyl Faced Drywall	.015	S.F.
Sound-deadening Board	.009	S.F.
Walls in Place		
2" x 4" Studs with 5/8"		
Gypsum Drywall Both Sides Taped	.053	S.F.
2" x 4" Studs with 2 Layers Gypsum Drywall		
Both Sides Taped	.078	S.F.

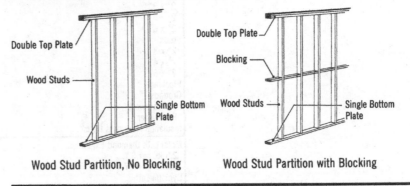

Wood Stud Partition, No Blocking

Wood Stud Partition with Blocking

Installation Time for Metal Lockers

Metal Lockers	Crew Makeup	Daily Output	Labor-Hours	Unit
LOCKERS Steel, baked enamel, 60" or 72", single tier				
Minimum	1 Sheet Metal Worker	14	.571	Opng.
Maximum		12	.667	Opng.
2 tier, 60" or 72" total height, minimum		26	.308	Opng.
Maximum		20	.400	Opng.
5 tier box lockers, minimum		30	.267	Opng.
Maximum		24	.333	Opng.
6 tier box lockers, minimum		36	.222	Opng.
Maximum		30	.267	Opng.
Basket rack with 32 baskets, 9" x 13" x 8" basket		50	.160	Basket
24 baskets, 12" x 13" x 8" basket		50	.160	Basket
Athletic, wire mesh, no lock, 18" x 18" x 72" basket	↓	12	.667	Ea.
Overhead locker baskets on chains, 14" x 14" baskets	3 Sheet Metal Workers	96	.250	Basket
Overhead locker framing system, add		600	.040	Basket
Locking rail and bench units, add	↓	120	.200	Basket
Locker bench, laminated maple, top only	1 Sheet Metal Worker	100	.080	L.F.
Pedestals, steel pipe	"	25	.320	Ea.
Teacher and pupil wardrobes, enameled				
22" x 15" x 61" high, minimum	1 Sheet Metal Worker	10	.800	Ea.
Average		9	.889	Ea.
Maximum		8	1.000	Ea.
Duplex lockers with 2 doors, 72" high, 15" x 15"		10	.800	Ea.
15" x 21"	↓	10	.800	Ea.

Pipe Sizing for Heating

Heating Load, BTU/HR.	GPM Circulated (20°T.D.)	Recommended Connecting Tubing Size (Type M) for Various Heating Loads and Connecting Tubing Lengths. (Figures based on 10,000 BTU per GPM, or on temperature drop of 20° thru the circuit.)				
		Total Length Ft. Connecting Tubing				
		0–50	50–100	100–150	150–200	200–300
		Tubing, Nominal O.D., Type M				
5,000	0.5	3/8	3/8	1/2	1/2	1/2
10,000	1.0	3/8	3/8	1/2	1/2	1/2
15,000	1.5	1/2	1/2	1/2	1/2	3/4
20,000	2.0	1/2	1/2	1/2	3/4	3/4
30,000	3.0	1/2	1/2	3/4	3/4	3/4
40,000	4.0	1/2	3/4	3/4	3/4	3/4
50,000	5.0	3/4	3/4	3/4	1	1
60,000	6.0	3/4	3/4	1	1	1
75,000	7.5	3/4	1	1	1	1
100,000	10.0	1	1	1	1-1/4	1-1/4
125,000	12.5	1	1	1-1/4	1-1/4	1-1/4
150,000	15.0	1	1-1/4	1-1/4	1-1/4	1-1/2
200,000	20.0	1-1/4	1-1/4	1-1/2	1-1/2	1-1/2
250,000	25.0	1-1/4	1-1/2	1-1/2	2	2
300,000	30.0	1-1/2	1-1/2	2	2	2
400,000	40.0	2	2	2	2	2
500,000	50.0	2	2	2	2-1/2	2-1/2
600,000	60.0	2	2	2-1/2	2-1/2	2-1/2
800,000	80.0	2-1/2	2-1/2	2-1/2	2-1/2	3
1,000,000	100.0	2-1/2	2-1/2	3	3	3
1,250,000	125.0	2-1/2	3	3	3	3-1/2
1,500,000	150.0	3	3	3	3-1/2	3-1/2
2,000,000	200.0	3	3-1/2	3-1/2	4	4
2,500,000	250.0	3-1/2	3-1/2	4	4	5
3,000,000	300.0	3-1/2	4	4	4	5
4,000,000	400.0	4	4	5	5	5
5,000,000	500.0	5	5	5	6	6
6,000,000	600.0	5	6	6	6	8
8,000,000	800.0	8	6	8	8	8
10,000,000	1,000.0	8	8	8	8	10

Installation Time in Labor-Hours for Plumbing Fixtures

Description	Labor-Hours	Unit
For Setting Fixture and Trim		
Bath Tub	3.636	Ea.
Bidet	3.200	Ea.
Dental Fountain	2.000	Ea.
Drinking Fountain	2.500	Ea.
Lavatory		
Vanity Top	2.500	Ea.
Wall Hung	2.000	Ea.
Laundry Sinks	2.667	Ea.
Prison/Institution Fixtures		
Lavatory	2.000	Ea.
Service Sink	5.333	Ea.
Urinal	4.000	Ea.
Water Closet	2.759	Ea.
Combination Water Closet and Lavatory	3.200	Ea.
Shower Stall	8.000	Ea.
Sinks		
Corrosion Resistant	5.333	Ea.
Kitchen, Countertop	3.330	Ea.
Kitchen, Raised Deck	7.270	Ea.
Service, Floor	3.640	Ea.
Service, Wall	4.000	Ea.
Urinals		
Wall Hung	5.333	Ea.
Stall Type	6.400	Ea.
Wash Fountain, Group	9.600	Ea.
Water Closets		
Tank Type, Wall Hung	3.019	Ea.
Floor Mount, One Piece	3.019	Ea.
Bowl Only, Wall Hung	2.759	Ea.
Bowl Only, Floor Mount	2.759	Ea.
Gang, Side by Side, First	2.759	Ea.
Each Additional	2.759	Ea.
Gang, Back to Back, First Pair	5.520	Pair
Each Additional Pair	5.520	Pair
Water Conserving Type	2.963	Ea.
Water Cooler	4.000	Ea.

Installation Time in Labor-Hours for Conduit

Description	Labor-Hours	Unit
Rigid Galvanized Steel 1/2" Diameter	.089	L.F.
1-1/2" Diameter	.145	L.F.
3" Diameter	.320	L.F.
6" Diameter	.800	L.F.
Aluminum 1/2" Diameter	.080	L.F.
1-1/2" Diameter	.123	L.F.
3" Diameter	.178	L.F.
6" Diameter	.400	L.F.
IMC 1/2" Diameter	.080	L.F.
1-1/2" Diameter	.133	L.F.
3" Diameter	.267	L.F.
4" Diameter	.320	L.F.
Plastic Coated Rigid Steel 1/2" Diameter	.100	L.F.
1-1/2" Diameter	.178	L.F.
3" Diameter	.364	L.F.
6" Diameter	.800	L.F.
EMT 1/2" Diameter	.047	L.F.
1-1/2" Diameter	.089	L.F.
3" Diameter	.160	L.F.
4" Diameter	.200	L.F.
PVC Nonmetallic 1/2" Diameter	.042	L.F.
1-1/2" Diameter	.080	L.F.
3" Diameter	.145	L.F.
6" Diameter	.267	L.F.

Rigid Steel, Plastic Coated Coupling

PVC Conduit

PVC Elbow

Aluminum Conduit

EMT Set Screw Connector

Aluminum Elbow

EMT Connector

Rigid Steel, Plastic Coated Conduit

EMT to Conduit Adapter

Rigid Steel, Plastic Coated Elbow

EMT to Greenfield Adapter

Installation Time in Labor-Hours for Wiring Devices

Description	Labor-Hours	Unit
Receptacle 20A 250V	.290	Ea.
Receptacle 30A 250V	.530	Ea.
Receptacle 50A 250V	.720	Ea.
Receptacle 60A 250V	1.000	Ea.
Box, 4" Square	.400	Ea.
Box, Single Gang	.290	Ea.
Box, Cast Single Gang	.660	Ea.
Cover, Weatherproof	.120	Ea.
Cover, Raised Device	.150	Ea.
Cover, Brushed Brass	.100	Ea.

30 Amp, 125 Volt, Nema 5

50 Amp, 125 Volt, Nema 5

20 Amp, 250 Volt, Nema 6

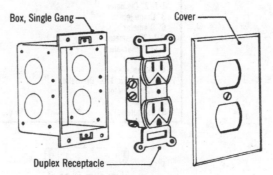

Box, Single Gang

Cover

Duplex Receptacle

Receptacle, Including Box and Cover

Receptacles

Appendix B

The following pages are a list of abbreviations from *Means Building Construction Cost Data*. They represent standard abbreviations used in Means data, as well as throughout the building construction industry, for the most common materials, units of measure, and associations.

A	Area Square Feet; Ampere	Cab.	Cabinet	Demob.	Demobilization
ABS	Acrylonitrile Butadiene Stryrene; Asbestos Bonded Steel	Cair.	Air Tool Laborer	d.f.u.	Drainage Fixture Units
A.C.	Alternating Current; Air-Conditioning; Asbestos Cement; Plywood Grade A & C	Calc	Calculated	D.H.	Double Hung
		Cap.	Capacity	DHW	Domestic Hot Water
		Carp.	Carpenter	Diag.	Diagonal
		C.B.	Circuit Breaker	Diam.	Diameter
		C.C.A.	Chromate Copper Arsenate	Distrib.	Distribution
A.C.I.	American Concrete Institute	C.C.F.	Hundred Cubic Feet	Dk.	Deck
AD	Plywood, Grade A & D	cd	Candela	D.L.	Dead Load; Diesel
Addit.	Additional	cd/sf	Candela per Square Foot	DLH	Deep Long Span Bar Joist
Adj.	Adjustable	CD	Grade of Plywood Face & Back	Do.	Ditto
af	Audio-frequency	CDX	Plywood, Grade C & D, exterior glue	Dp.	Depth
A.G.A.	American Gas Association			D.P.S.T.	Double Pole, Single Throw
Agg.	Aggregate	Cefi.	Cement Finisher	Dr.	Driver
A.H.	Ampere Hours	Cem.	Cement	Drink.	Drinking
A hr.	Ampere-hour	CF	Hundred Feet	D.S.	Double Strength
A.H.U.	Air Handling Unit	C.F.	Cubic Feet	D.S.A.	Double Strength A Grade
A.I.A.	American Institute of Architects	CFM	Cubic Feet per Minute	D.S.B.	Double Strength B Grade
AIC	Ampere Interrupting Capacity	c.g.	Center of Gravity	Dty.	Duty
Allow.	Allowance	CHW	Chilled Water; Commercial Hot Water	DWV	Drain Waste Vent
alt.	Altitude			DX	Deluxe White, Direct Expansion
Alum.	Aluminum	C.I.	Cast Iron	dyn	Dyne
a.m.	Ante Meridiem	C.I.P.	Cast in Place	e	Eccentricity
Amp.	Ampere	Circ.	Circuit	E	Equipment Only; East
Anod.	Anodized	C.L.	Carload Lot	Ea.	Each
Approx.	Approximate	Clab.	Common Laborer	E.B.	Encased Burial
Apt.	Apartment	Clam	Common maintenance laborer	Econ.	Economy
Asb.	Asbestos	C.L.F.	Hundred Linear Feet	E.C.Y	Embankment Cubic Yards
A.S.B.C.	American Standard Building Code	CLF	Current Limiting Fuse	EDP	Electronic Data Processing
Asbe.	Asbestos Worker	CLP	Cross Linked Polyethylene	EIFS	Exterior Insulation Finish System
A.S.H.R.A.E.	American Society of Heating, Refrig. & AC Engineers	cm	Centimeter	E.D.R.	Equiv. Direct Radiation
		CMP	Corr. Metal Pipe	Eq.	Equation
A.S.M.E.	American Society of Mechanical Engineers	C.M.U.	Concrete Masonry Unit	Elec.	Electrician; Electrical
		CN	Change Notice	Elev.	Elevator; Elevating
A.S.T.M.	American Society for Testing and Materials	Col.	Column	EMT	Electrical Metallic Conduit; Thin Wall Conduit
		CO_2	Carbon Dioxide		
Attchmt.	Attachment	Comb.	Combination	Eng.	Engine, Engineered
Avg.	Average	Compr.	Compressor	EPDM	Ethylene Propylene Diene Monomer
A.W.G.	American Wire Gauge	Conc.	Concrete		
AWWA	American Water Works Assoc.	Cont.	Continuous; Continued	EPS	Expanded Polystyrene
Bbl.	Barrel	Corr.	Corrugated	Eqhv.	Equip. Oper., Heavy
B&B	Grade B and Better; Balled & Burlapped	Cos	Cosine	Eqlt.	Equip. Oper., Light
		Cot	Cotangent	Eqmd.	Equip. Oper., Medium
B.&S.	Bell and Spigot	Cov.	Cover	Eqmm.	Equip. Oper., Master Mechanic
B.&W.	Black and White	C/P	Cedar on Paneling	Eqol.	Equip. Oper., Oilers
b.c.c.	Body-centered Cubic	CPA	Control Point Adjustment	Equip.	Equipment
B.C.Y.	Bank Cubic Yards	Cplg.	Coupling	ERW	Electric Resistance Welded
BE	Bevel End	C.P.M.	Critical Path Method	E.S.	Energy Saver
B.F.	Board Feet	CPVC	Chlorinated Polyvinyl Chloride	Est.	Estimated
Bg. cem.	Bag of Cement	C.Pr.	Hundred Pair	esu	Electrostatic Units
BHP	Boiler Horsepower; Brake Horsepower	CRC	Cold Rolled Channel	E.W.	Each Way
		Creos.	Creosote	EWT	Entering Water Temperature
B.I.	Black Iron	Crpt.	Carpet & Linoleum Layer	Excav.	Excavation
Bit.; Bitum.	Bituminous	CRT	Cathode-ray Tube	Exp.	Expansion, Exposure
Bk.	Backed	CS	Carbon Steel, Constant Shear Bar Joist	Ext.	Exterior
Bkrs.	Breakers			Extru.	Extrusion
Bldg.	Building	Csc	Cosecant	f.	Fiber stress
Blk.	Block	C.S.F.	Hundred Square Feet	F	Fahrenheit; Female; Fill
Bm.	Beam	CSI	Construction Specifications Institute	Fab.	Fabricated
Boil.	Boilermaker			FBGS	Fiberglass
B.P.M.	Blows per Minute	C.T.	Current Transformer	F.C.	Footcandles
BR	Bedroom	CTS	Copper Tube Size	f.c.c.	Face-centered Cubic
Brg.	Bearing	Cu	Copper, Cubic	f'c.	Compressive Stress in Concrete; Extreme Compressive Stress
Brhe.	Bricklayer Helper	Cu. Ft.	Cubic Foot		
Bric.	Bricklayer	cw	Continuous Wave	F.E.	Front End
Brk.	Brick	C.W.	Cool White; Cold Water	FEP	Fluorinated Ethylene Propylene (Teflon)
Brng.	Bearing	Cwt.	100 Pounds		
Brs.	Brass	C.W.X.	Cool White Deluxe	F.G.	Flat Grain
Brz.	Bronze	C.Y.	Cubic Yard (27 cubic feet)	F.H.A.	Federal Housing Administration
Bsn.	Basin	C.Y./Hr.	Cubic Yard per Hour	Fig.	Figure
Btr.	Better	Cyl.	Cylinder	Fin.	Finished
BTU	British Thermal Unit	d	Penny (nail size)	Fixt.	Fixture
BTUH	BTU per Hour	D	Deep; Depth; Discharge	Fl. Oz.	Fluid Ounces
B.U.R.	Built-up Roofing	Dis.;Disch.	Discharge	Flr.	Floor
BX	Interlocked Armored Cable	Db.	Decibel	F.M.	Frequency Modulation; Factory Mutual
c	Conductivity, Copper Sweat	Dbl.	Double		
C	Hundred; Centigrade	DC	Direct Current	Fmg.	Framing
C/C	Center to Center, Cedar on Cedar	DDC	Direct Digital Control	Fndtn.	Foundation

Fori.	Foreman, Inside	I.W.	Indirect Waste	M.C.F.	Thousand Cubic Feet
Foro.	Foreman, Outside	J	Joule	M.C.F.M.	Thousand Cubic Feet per Minute
Fount.	Fountain	J.I.C.	Joint Industrial Council	M.C.M.	Thousand Circular Mils
FPM	Feet per Minute	K	Thousand; Thousand Pounds;	M.C.P.	Motor Circuit Protector
FPT	Female Pipe Thread		Heavy Wall Copper Tubing, Kelvin	MD	Medium Duty
Fr.	Frame	K.A.H.	Thousand Amp. Hours	M.D.O.	Medium Density Overlaid
F.R.	Fire Rating	KCMIL	Thousand Circular Mils	Med.	Medium
FRK	Foil Reinforced Kraft	KD	Knock Down	MF	Thousand Feet
FRP	Fiberglass Reinforced Plastic	K.D.A.T.	Kiln Dried After Treatment	M.F.B.M.	Thousand Feet Board Measure
FS	Forged Steel	kg	Kilogram	Mfg.	Manufacturing
FSC	Cast Body; Cast Switch Box	kG	Kilogauss	Mfrs.	Manufacturers
Ft.	Foot; Feet	kgf	Kilogram Force	mg	Milligram
Ftng.	Fitting	kHz	Kilohertz	MGD	Million Gallons per Day
Ftg.	Footing	Kip.	1000 Pounds	MGPH	Thousand Gallons per Hour
Ft. Lb.	Foot Pound	KJ	Kiljoule	MH, M.H.	Manhole; Metal Halide; Man-Hour
Furn.	Furniture	K.L.	Effective Length Factor	MHz	Megahertz
FVNR	Full Voltage Non-Reversing	K.L.F	Kips per Linear Foot	Mi.	Mile
FXM	Female by Male	Km	Kilometer	MI	Malleable Iron; Mineral Insulated
Fy.	Minimum Yield Stress of Steel	K.S.F.	Kips per Square Foot	mm	Millimeter
g	Gram	K.S.I.	Kips per Square Inch	Mill.	Millwright
G	Gauss	kV	Kilovolt	Min., min.	Minimum, minute
Ga.	Gauge	kVA	Kilovolt Ampere	Misc.	Miscellaneous
Gal.	Gallon	K.V.A.R.	Kilovar (Reactance)	ml	Milliliter, Mainline
Gal./Min.	Gallon per Minute	KW	Kilowatt	M.L.F.	Thousand Linear Feet
Galv.	Galvanized	KWh	Kilowatt-hour	Mo.	Month
Gen.	General	L	Labor Only; Length; Long;	Mobil.	Mobilization
G.F.I.	Ground Fault Interrupter		Medium Wall Copper Tubing	Mog.	Mogul Base
Glaz.	Glazier	Lab.	Labor	MPH	Miles per Hour
GPD	Gallons per Day	lat	Latitude	MPT	Male Pipe Thread
GPH	Gallons per Hour	Lath.	Lather	MRT	Mile Round Trip
GPM	Gallons per Minute	Lav.	Lavatory	ms	Millisecond
GR	Grade	lb.; #	Pound	M.S.F.	Thousand Square Feet
Gran.	Granular	L.B.	Load Bearing; L Conduit Body	Mstz.	Mosaic & Terrazzo Worker
Grnd.	Ground	L. & E.	Labor & Equipment	M.S.Y.	Thousand Square Yards
H	High; High Strength Bar Joist;	lb./hr.	Pounds per Hour	Mtd.	Mounted
	Henry	lb./L.F.	Pounds per Linear Foot	Mthe.	Mosaic & Terrazzo Helper
H.C.	High Capacity	lbf/sq.in.	Pound-force per Square Inch	Mtng.	Mounting
H.D.	Heavy Duty; High Density	L.C.L.	Less than Carload Lot	Mult.	Multi; Multiply
H.D.O.	High Density Overlaid	L.C.Y.	Loose Cubic Yard	M.V.A.	Million Volt Amperes
Hdr.	Header	Ld.	Load	M.V.A.R.	Million Volt Amperes Reactance
Hdwe.	Hardware	LE	Lead Equivalent	MV	Megavolt
Help.	Helper Average	LED	Light Emitting Diode	MW	Megawatt
HEPA	High Efficiency Particulate Air	L.F.	Linear Foot	MXM	Male by Male
	Filter	Lg.	Long; Length; Large	MYD	Thousand Yards
Hg	Mercury	L & H	Light and Heat	N	Natural; North
HIC	High Interrupting Capacity	LH	Long Span Bar Joist	nA	Nanoampere
HM	Hollow Metal	L.H.	Labor Hours	NA	Not Available; Not Applicable
H.O.	High Output	L.L.	Live Load	N.B.C.	National Building Code
Horiz.	Horizontal	L.L.D.	Lamp Lumen Depreciation	NC	Normally Closed
H.P.	Horsepower; High Pressure	lm	Lumen	N.E.M.A.	National Electrical Manufacturers
H.P.F.	High Power Factor	lm/sf	Lumen per Square Foot		Assoc.
Hr.	Hour	lm/W	Lumen per Watt	NEHB	Bolted Circuit Breaker to 600V.
Hrs./Day	Hours per Day	L.O.A.	Length Over All	N.L.B.	Non-Load-Bearing
HSC	High Short Circuit	log	Logarithm	NM	Non-Metallic Cable
Ht.	Height	L-O-L	Lateralolet	nm	Nanometer
Htg.	Heating	L.P.	Liquefied Petroleum; Low Pressure	No.	Number
Htrs.	Heaters	L.P.F.	Low Power Factor	NO	Normally Open
HVAC	Heating, Ventilation & Air-	LR	Long Radius	N.O.C.	Not Otherwise Classified
	Conditioning	L.S.	Lump Sum	Nose.	Nosing
Hvy.	Heavy	Lt.	Light	N.P.T.	National Pipe Thread
HW	Hot Water	Lt. Ga.	Light Gauge	NQOD	Combination Plug-on/Bolt on
Hyd.;Hydr.	Hydraulic	L.T.L.	Less than Truckload Lot		Circuit Breaker to 240V.
Hz.	Hertz (cycles)	Lt. Wt.	Lightweight	N.R.C.	Noise Reduction Coefficient
I.	Moment of Inertia	L.V.	Low Voltage	N.R.S.	Non Rising Stem
I.C.	Interrupting Capacity	M	Thousand; Material; Male;	ns	Nanosecond
ID	Inside Diameter		Light Wall Copper Tubing	nW	Nanowatt
I.D.	Inside Dimension; Identification	M²CA	Meters Squared Contact Area	OB	Opposing Blade
I.F.	Inside Frosted	m/hr; M.H.	Man-hour	OC	On Center
I.M.C.	Intermediate Metal Conduit	mA	Milliampere	OD	Outside Diameter
In.	Inch	Mach.	Machine	O.D.	Outside Dimension
Incan.	Incandescent	Mag. Str.	Magnetic Starter	ODS	Overhead Distribution System
Incl.	Included; Including	Maint.	Maintenance	O.G.	Ogee
Int.	Interior	Marb.	Marble Setter	O.H.	Overhead
Inst.	Installation	Mat; Mat'l.	Material	O&P	Overhead and Profit
Insul.	Insulation/Insulated	Max.	Maximum	Oper.	Operator
I.P.	Iron Pipe	MBF	Thousand Board Feet	Opng.	Opening
I.P.S.	Iron Pipe Size	MBH	Thousand BTU's per hr.	Orna.	Ornamental
I.P.T.	Iron Pipe Threaded	MC	Metal Clad Cable	OSB	Oriented Strand Board

O.S.&Y.	Outside Screw and Yoke	Rsr	Riser	Th.;Thk.	Thick
Ovhd.	Overhead	RT	Round Trip	Thn.	Thin
OWG	Oil, Water or Gas	S.	Suction; Single Entrance; South	Thrded	Threaded
Oz.	Ounce	SC	Screw Cover	Tilf.	Tile Layer, Floor
P.	Pole; Applied Load; Projection	SCFM	Standard Cubic Feet per Minute	Tilh.	Tile Layer, Helper
p.	Page	Scaf.	Scaffold	THHN	Nylon Jacketed Wire
Pape.	Paperhanger	Sch.; Sched.	Schedule	THW.	Insulated Strand Wire
P.A.P.R.	Powered Air Purifying Respirator	S.C.R.	Modular Brick	THWN;	Nylon Jacketed Wire
PAR	Parabolic Reflector	S.D.	Sound Deadening	T.L.	Truckload
Pc., Pcs.	Piece, Pieces	S.D.R.	Standard Dimension Ratio	T.M.	Track Mounted
P.C.	Portland Cement; Power Connector	S.E.	Surfaced Edge	Tot.	Total
P.C.F.	Pounds per Cubic Foot	Sel.	Select	T-O-L	Threadolet
P.C.M.	Phase Contrast Microscopy	S.E.R.; S.E.U.	Service Entrance Cable	T.S.	Trigger Start
P.E.	Professional Engineer;	S.F.	Square Foot	Tr.	Trade
	Porcelain Enamel;	S.F.C.A.	Square Foot Contact Area	Transf.	Transformer
	Polyethylene; Plain End	S.F. Flr.	Square Foot of Floor	Trhv.	Truck Driver, Heavy
Perf.	Perforated	S.F.G.	Square Foot of Ground	Trlr	Trailer
Ph.	Phase	S.F. Hor.	Square Foot Horizontal	Trlt.	Truck Driver, Light
P.I.	Pressure Injected	S.F.R.	Square Feet of Radiation	TTY	Teletypewriter
Pile.	Pile Driver	S.F. Shlf.	Square Foot of Shelf	TV	Television
Pkg.	Package	S4S	Surface 4 Sides	T.W.	Thermoplastic Water Resistant
Pl.	Plate	Shee.	Sheet Metal Worker		Wire
Plah.	Plasterer Helper	Sin.	Sine	UCI	Uniform Construction Index
Plas.	Plasterer	Skwk.	Skilled Worker	UF	Underground Feeder
Pluh.	Plumbers Helper	SL	Saran Lined	UGND	Underground Feeder
Plum.	Plumber	S.L.	Slimline	U.H.F.	Ultra High Frequency
Ply.	Plywood	Sldr.	Solder	U.L.	Underwriters Laboratory
p.m.	Post Meridiem	SLH	Super Long Span Bar Joist	Unfin.	Unfinished
Pntd.	Painted	S.N.	Solid Neutral	URD	Underground Residential
Pord.	Painter, Ordinary	S-O-L	Socketolet		Distribution
pp	Pages	sp	Standpipe	US	United States
PP; PPL	Polypropylene	S.P.	Static Pressure; Single Pole; Self-	USP	United States Primed
P.P.M.	Parts per Million		Propelled	UTP	Unshielded Twisted Pair
Pr.	Pair	Spri.	Sprinkler Installer	V	Volt
P.E.S.B.	Pre-engineered Steel Building	spwg	Static Pressure Water Gauge	V.A.	Volt Amperes
Prefab.	Prefabricated	S.P.D.T.	Single Pole, Double Throw	V.C.T.	Vinyl Composition Tile
Prefin.	Prefinished	SPF	Spruce Pine Fir	VAV	Variable Air Volume
Prop.	Propelled	S.P.S.T.	Single Pole, Single Throw	VC	Veneer Core
PSF; psf	Pounds per Square Foot	SPT	Standard Pipe Thread	Vent.	Ventilation
PSI; psi	Pounds per Square Inch	Sq.	Square; 100 Square Feet	Vert.	Vertical
PSIG	Pounds per Square Inch Gauge	Sq. Hd.	Square Head	V.F.	Vinyl Faced
PSP	Plastic Sewer Pipe	Sq. In.	Square Inch	V.G.	Vertical Grain
Pspr.	Painter, Spray	S.S.	Single Strength; Stainless Steel	V.H.F.	Very High Frequency
Psst.	Painter, Structural Steel	S.S.B.	Single Strength B Grade	VHO	Very High Output
P.T.	Potential Transformer	sst	Stainless Steel	Vib.	Vibrating
P. & T.	Pressure & Temperature	Sswk.	Structural Steel Worker	V.L.F.	Vertical Linear Foot
Ptd.	Painted	Sswl.	Structural Steel Welder	Vol.	Volume
Ptns.	Partitions	St.;Stl.	Steel	VRP	Vinyl Reinforced Polyester
Pu	Ultimate Load	S.T.C.	Sound Transmission Coefficient	W	Wire; Watt; Wide; West
PVC	Polyvinyl Chloride	Std.	Standard	w/	With
Pvmt.	Pavement	STK	Select Tight Knot	W.C.	Water Column; Water Closet
Pwr.	Power	STP	Standard Temperature & Pressure	W.F.	Wide Flange
Q	Quantity Heat Flow	Stpi.	Steamfitter, Pipefitter	W.G.	Water Gauge
Quan.;Qty.	Quantity	Str.	Strength; Starter; Straight	Wldg.	Welding
Q.C.	Quick Coupling	Strd.	Stranded	W. Mile	Wire Mile
r	Radius of Gyration	Struct.	Structural	W-O-L	Weldolet
R	Resistance	Sty.	Story	W.R.	Water Resistant
R.C.P.	Reinforced Concrete Pipe	Subj.	Subject	Wrck.	Wrecker
Rect.	Rectangle	Subs.	Subcontractors	W.S.P.	Water, Steam, Petroleum
Reg.	Regular	Surf.	Surface	WT., Wt.	Weight
Reinf.	Reinforced	Sw.	Switch	WWF	Welded Wire Fabric
Req'd.	Required	Swbd.	Switchboard	XFER	Transfer
Res.	Resistant	S.Y.	Square Yard	XFMR	Transformer
Resi.	Residential	Syn.	Synthetic	XHD	Extra Heavy Duty
Rgh.	Rough	S.Y.P.	Southern Yellow Pine	XHHW; XLPE	Cross-Linked Polyethylene Wire
RGS	Rigid Galvanized Steel	Sys.	System		Insulation
R.H.W.	Rubber, Heat & Water Resistant;	t.	Thickness	XLP	Cross-linked Polyethylene
	Residential Hot Water	T	Temperature; Ton	Y	Wye
rms	Root Mean Square	Tan	Tangent	yd	Yard
Rnd.	Round	T.C.	Terra Cotta	yr	Year
Rodm.	Rodman	T & C	Threaded and Coupled	Δ	Delta
Rofc.	Roofer, Composition	T.D.	Temperature Difference	%	Percent
Rofp.	Roofer, Precast	Tdd	Telecommunications Device for	~	Approximately
Rohe.	Roofer Helpers (Composition)		the Deaf	Ø	Phase
Rots.	Roofer, Tile & Slate	T.E.M.	Transmission Electron Microscopy	@	At
R.O.W.	Right of Way	TFE	Tetrafluoroethylene (Teflon)	#	Pound; Number
RPM	Revolutions per Minute	T. & G.	Tongue & Groove;	<	Less Than
R.S.	Rapid Start		Tar & Gravel	>	Greater Than

Notes

Notes

Notes

416

Notes

Notes